Bitcoin for Nonmathematicians

Exploring the Foundations of Crypto Payments

Slava Gomzin

Universal-Publishers
Boca Raton

Bitcoin for Nonmathematicians: Exploring the Foundations of Crypto Payments

Universal-Publishers
Boca Raton, Florida • USA
2016

ISBN-10: 1-62734-071-8
ISBN-13: 978-1-62734-071-7

www.universal-publishers.com

Publisher's Cataloging-in-Publication Data

Names: Gomzin, Slava.
Title: Bitcoin for nonmathematicians : exploring the foundations of
    crypto payments / Slava Gomzin.
Description: Boca Raton, FL : Universal Publishers, 2016. | Includes
    bibliographical references and index.
Identifiers: LCCN 2016930001 | ISBN 978-1-62734-071-7 (pbk.)
Subjects: LCSH: Bitcoin. | Money. | Electronic commerce. | Mobile
    commerce. | Cryptography--Data processing. | Data encryption
    (Computer science) | BISAC: BUSINESS & ECONOMICS /
    Money & Monetary Policy. | BUSINESS & ECONOMICS / E-
    Commerce / General. | COMPUTERS / Electronic Commerce.
    | COMPUTERS / Security / Cryptography.
Classification: LCC HF5548.32 .G659 2016 (print) | DDC: 332.4--
    dc23.

To Svetlana
and our daughters Alona, Aliza, and Arina

About the Author

 Slava Gomzin is Director of Information Security at PCCI (Parkland Center for Clinical Innovation), a nonprofit research and development corporation delivering real time predictive analytics solutions. Slava is also the author of *Hacking Point of Sale: Payment Application Secrets, Threats, and Solutions* (Wiley, 2014), and has written many articles on payment security and technology. Prior to joining PCCI, Slava was a security and payments technologist at Hewlett-Packard, where he helped create products that are integrated into modern payment processing ecosystems. Before HP, he worked as a security architect, corporate product security officer, and R&D and application security manager at Retalix, a division of NCR Retail. As PCI ISA, he focused on security and PA-DSS, PCI DSS, and PCI P2PE compliance of POS systems, payment applications, and gateways. Slava currently holds CISSP, PCIP, ECSP, and Security+ certifications. He blogs about information security and technology at www.gomzin.com.

Credits

Technical Editor
Ken Westin

Copy Editor
Adaobi Obi Tutton

Foreword
Doug McClellan

Publisher & CEO
Jeff Young

Photo
Svetlana Gomzin

Production Editor
Christie Mayer

Cover Design
Ivan Popov

Acknowledgments

Writing a book is not easy and cannot succeed without help from other people. First of all, I would like to thank Carol Long for convincing me to start writing another book right after the previous one was published. And thanks to Jeff Young for bringing this project to reality. Also, I would like to thank my ex-coworkers from HP, especially David White for his support and interest in such a controversial topic. Thanks to Ken Westin for his enthusiastic support and contribution. Thanks also go to VentureBeat editor, Morwenna Marshall, for the opportunity to share my ideas with a wider audience. Thanks to Adaobi Obi Tulton for another great editorial effort. Special thanks to Doug McClellan for his bright and sincere foreword. And finally, I want to thank my wife, Svetlana, for her continuous support and understanding.

Contents at a Glance

Contents

Foreword

by Doug McClellan

I'm a numismatist, which is a fancy word for a coin collector. I'm also a software developer for electronic funds transfer (EFT) systems by profession and, like most readers here, also an investor.

So when Slava told me he was writing a book about bitcoins, I knew I wanted to read it because it was an area I've always had an interest in from the three perspectives I've just mentioned. Bitcoins mainly tie into the future of electronic payments, but also have been used as an investment vehicle and could very well have an impact on the future of numismatics.

I've been collecting coins ever since I was a kid, and started building my collection over 45 years ago with a Lincoln Cent album. For the last 25 years I've developed software for the retail merchant industry, and specialized in EFT systems for the convenience store market segment for the last 17 years. When you buy a soda at the convenience store or swipe your card at the pump, there is software needed to process your transaction electronically.

I will expand more on that in a bit, but first I want to introduce you to the author, Slava Gomzin. For those of you who are not familiar with his work from his blog at www.gomzin.com or from the other books he has published in the area of cybersecurity, such as *Hacking Point of Sale*, along with his *Application Security* and *Cyber Privacy* book series titles for electronic data security, I think you will join me in appreciating his insight in this area.

I met Slava in 1999 when we worked together to create an EFT software system through our mutual employer. Slava had emigrated from Russia to Israel when President Reagan had challenged the Russian government to allow its citizens to have more freedom in their lives. Slava was one of those people who saw the opportunity and had the courage to build a new life in a foreign country. He moved his family to Israel, where he found employment using his computer programming skills. Later he again utilized his pioneer spirit when he moved with his wife and children to America, the true land of opportunity.

Slava has proven that hard work and dedication, along with natural talent and abilities, will flourish in a free society. Slava was our team leader in the EFT development group during a time when our company was rapidly expanding here in the United States. While managing multiple development projects with different EFT networks, he had taken an interest in cyber security, which was in its infancy at the time. He read, studied, and attended courses in cybersecurity, and has earned many certifications over the years. Slava also served on the PCI standards committee when the early standards were being developed. So, as you can see, Slava knows cybersecurity. In fact, I would say he is an expert in the field.

In this book, Slava brings the reader along on a journey from the origins of money and electronic payments and into the implementation of bitcoins as a cybercurrency.

I find the term bitcoin to be rather clever as a name. It is not, of course, a coin in the physical sense, but an electronic implementation of money represented by bits, the electronic 1s and 0s that computers use to store data. The origin of bitcoin is rather mysterious, as you will learn in the book.

Using the standard economic concepts that the value of anything is what a willing buyer will pay a willing seller in an arm's length transaction, cost is what you give up to get something else, and money is a standardization of trade units that allow for marketplace transactions to occur, bitcoins are an attempt to create a new type of currency that is separate from a central system (such as government-issued currency) and that can also be deployed as an electronic payment system.

Throughout history, money has always been physical. The earliest coinage originated in Asia Minor about 2,500 years ago from an alloy known as electrum or "elektron" to the Greeks. It is composed of silver and gold, along with other trace metals, occurs naturally in nugget form, and is found in riverbeds. It worked well for its purpose prior to the development of technology needed to separate elements. Merchants allowed trusted customers to carry a tab (the first use of credit) and pay with electrum coins when the bill was sufficiently high. The nuggets varied in size and weight and were treated as bullion. The first designs on coins were simple striation lines, which mimicked the lines formed on the nuggets from the water flow in rivers. It was Aristotle that championed the importance of having an image on the obverse, which really transitioned bullion into true coinage.

In early colonial America, daily commerce was conducted using coins produced by the official mints of other established nations, along with a hodgepodge of tokens and medals issued by private individuals and mints from inside and outside of America. The first coins issued by the authority of the United States were the Fugio pieces in 1787, and they are some of my personal favorite coins. The design had 13 interlocking circles and a small circle in the middle with the words "United States" around it and the words "We Are One" in the center. On the other side there was a sundial with a meridian Sun above it, the word "Fugio" (the intended meaning is time flies) on the left, and the year 1787 to the right of the sundial. Under the sundial are the words "Mind Your Business," a saying credited to Benjamin Franklin. To me, this coin encompasses a lot of pride, solidarity, and hope for the young United States of America.

An important characteristic of a sovereign nation is the right to issue its own coins, and America began exercising that right in 1792 by issuing pattern coins, followed by copper coins in 1793, silver coins in 1794, and gold coins in 1795. Before the denominations we have circulating today, there have been some more unusual ones, starting with the half cent in 1793, two-cent pieces (1864–1873) in which the motto "In God We Trust" first appeared, along with three-cent pieces (1851–1889). There have also been half dimes (1794–1873) and twenty-cent pieces (1875–1878). Gold coins have been minted in denominations of $1, $2.50, $3, $4, $5, $10, and $20. Gold $50 and platinum $100 coins are issued today by the US Mint, but these are considered bullion. There have been various reasons for the different denominations, but bitcoin transactions can occur in fractions of a bitcoin, making them very versatile.

As our society moves to a cashless environment, I wonder how that will impact future coin collectors. Bitcoins will never become a collectable, since they lack the characteristics of physical coins. Blockchains are free to anyone and have no varying condition state from circulating. At some future point in time, there won't be a need for physical coinage and the billions of coins the US Mint currently produces each year will become obsolete. Will there still be an interest in collecting something that future generations would have never used for their intended purpose in their daily lives? Only time will tell.

The future of bitcoins is also unknown. Early investors had a wild ride with large gains followed by large declines as they sought to find bitcoins' true value in relation to other currencies. They had started to obtain a reputation as taboo due to their use in criminal activities based on the notion that they can be held anonymously. But as Slava explains, bitcoins are not entirely anonymous and can be traced and tracked back to a unique IP address.

One thing is certain: bitcoins are becoming mainstream, and with their lower cost as a payment system, many merchants not only accept bitcoins as tender, some actually prefer them as a cost-saving method for processing electronic payments.

As you read this book, you will learn both the history and possible future of bitcoins. With Slava's in-depth analysis of the security aspect of bitcoin financial transactions, perhaps you will learn to prefer this cryptocurrency system as well.

Introduction

There are no conditions of life to which a man cannot get
accustomed, especially if he sees them accepted by everyone
about him.
—Leo Tolstoy

Several years ago I was fascinated by an experiment I did. I was trying to live cashless, paying only with plastic cards, either debit or credit. My attempt was pretty successful until I went on a business trip abroad. My first (but not last!) failure was in a restaurant, when I received a check without a placeholder for a tip amount. There were no problems paying with a credit card, but there was no way to add a tip to the bill. So I had to ask my friend (who was not participating in my experiment) to pay a cash tip. The payment system, even though it was "aware" of electronic payments, was not fully integrated into the world of plastic money. Such a situation is still common in many places, especially outside North America and Europe.

I would face similar challenges today if suddenly I decided to do the same experiment with bitcoin, but this time the limitations would be different. Instead of geographical borders that divide the world into cash and cashless zones, there is an invisible Rubicon between the offline and online worlds. In this new version of my experiment, I could live a sustainable life without cash (or plastic) if I didn't leave my house. I could shop online and even order food from local restaurants. Whenever I needed to make a transfer of traditional money, for example, to pay the commodity bills (still virtual but counted in dollars rather bitcoins), I could exchange my bitcoins online and convert them to dollar transfer. I could even earn a living by mining the cryptocurrencies at home. However, this pattern breaks very quickly when you go offline and enter traditional brick-and-mortar stores. Few retailers today accept bitcoin or any other cryptocurrency, despite the obvious benefits: convenience, security, lower transaction fees, and attracting new generation of customers.

One of the most important goals of this book is to help people who are not closely familiar with math and cryptography to understand crypto payments. In order to do it smoothly and wisely, we need to understand several things, the first being the place cryptocurrency has in the modern payment ecosystem.

Don't let the fact that this book is technical scare you if you are not a programmer. This book can still be read by anyone who wants to get paid or pay with cryptocurrency, and the first several chapters will prove it by answering very basic questions, such as what are the players in the existing electronic payments game, and whether it is possible to integrate bitcoin into it painlessly without breaking the major rules.

While I realize that the readers of this book might be in a sense obsessed with crypto payments, we should stay calm and remember that there were (and in fact still are!) other types of currencies and methods of payment. Although bitcoin enthusiasts often use the term "revolution," from many perspectives, especially from the merchants' point of view, creation of cryptocurrency is just an evolution of a payment system that was made possible by modern science and technology, namely cryptography and the Internet.

If you ask how to characterize bitcoin in a single word, many would answer "cryptography." Although I agree with this answer, it is too generic, so my answer would be more specific (but contain more words): "public-key encryption and hash function." Here is why.

If we analyze existing payment systems—predecessors of bitcoin—there are two main problems in their design: security and centralization. Security flaws in the design of payment cards resulted in the creation of PCI data security standards, which forced merchants, service providers, banks, and payment brands to invest billions of dollars into security controls, which eventually failed to protect them from data breaches. On the other hand, as you will see in part I of this book, centralized management of the first virtual currencies was the main reason for fiasco.

Bitcoin design provides solutions to both the security and centralization problems: *digital signature* and *proof of work*. A digital signature is based on public-key cryptography, while a cryptographic hash function is the essential part of both a digital signature and a proof-of-work implementation.

Before the invention of digital signatures, it was impossible to broadcast the message throughout a public channel such as the

Internet and verify through multiple recipients that this message was unchanged since its creation by the original sender. Along with public-key encryption, the cryptographic hash function made creation of a digital signature possible, which protects the *integrity of crypto transactions*—a solution for *security problems*.

At the same time, a cryptographic one-way hash function, besides its participation in digital signature design, made proof-of-work implementation possible, which is a solution for *centralization problems*.

So it's safe to say that if you understand the cryptography behind bitcoin, then you know how bitcoin and other cryptocurrencies work, so you can trust them.

From Coins to Crypto

In This Part

CHAPTER 1

Traditional Money

Money is like muck, not good except it be spread
—Francis Bacon

Many books start telling their stories ab ovo,[1] and this book is no exception. One can say that bitcoins are created from thin air. While perhaps it's true, the idea wasn't born in a vacuum. There was life before bitcoin, and its daring ancestors helped to build a foundation for what we recognize today as cryptocurrency. Learning more about the history of traditional and digital money and payment methods might help us to understand the common rules of the game and predict the challenges facing cryptocurrencies.

This book does not pretend to be a full reference on bitcoin design and implementation, but rather focuses on cryptography. However, it would be reckless to discuss other aspects of cryptocurrencies without reviewing its implementation, and especially without comparing this design with other cashless payment implementations, such as plastic cards.

Commodities versus Gold

Money. Everyone knows what it is, at least its practical application. It all began with barter, when people were exchanging their products with each other for goods and services. For a very long time barter was the only way to sell or buy. But at some point, people realized that barter was limited and inconvenient. For example, I have oranges that I want to sell, and I need to buy some apples. But the apple seller doesn't need oranges, so I can't buy his apples. In this situation, I need to exchange my oranges for something that would satisfy the apple seller as a medium of exchange for his apples. This something can be a commodity—goods that are useful to many people and that can be easily and willingly swapped for other goods and services. So commodities became the first money. For some time, many different societies were happy using commodities, such as cocoa beans in Abyssinia or iron nails in Scotland, as money.[2] However, there were problems.

There are several important criteria for choosing a commodity to become money.[3] First, the commodity should be in widespread use, or in heavy demand, so everyone would be willing to accept it as a payment for their goods. Second, it should be highly divisible so that it can be divided into small chunks in order to process micropayments, for example. Another important feature is portability, which, in most cases, is boiled down to the high value per unit weight. If money units with relative low value are too heavy, it is difficult to carry and transport them. In addition, a commodity must be highly durable so that it can be reused many times and stored for a long time as savings. Due to this last requirement, foodstuffs, for example, cannot be used as money. Another property, fungibility, means that different pieces of the material are equal, and can be equally interchanged. For example, pure gold is fungible, while coffee beans are not because they can be a different type, age, quality, weight, and so on. And finally, the commodity must have a limited supply in order to maintain its value, meaning that there should be no easy way to voluntarily add large amounts of money to the existing money turnover. Gold, silver, and other precious metals are rare elements and thus they ideally fit this requirement.

Table 1-1: Conditions for Accepting Commodity as Money

	COCOA BEANS	SALT	GOLD	BITCOIN
Widespread use	-	●	●	-
Heavy demand (liquidity)	-	-	●	-
Highly divisible	-	●	●	●
High Value per Unit Weight	-	-	●	●
Easily Portable	-	-	●	●
Highly Durable	-	-	●	●
Fungible	-	-	●	●
Limited Supply	-	-	●	●

As you can see in Table 1-1, Bitcoin has much more of a chance to be accepted as money than coffee beans or salt; however, it still loses to gold, which is desirable by everyone due to historical traditions and physical characteristics.

Payment Cards

Electronic payments were introduced in the middle of twentieth century with the invention of magnetic stripe cards. Credit cards were the first application of plastic payment cards followed by various other types, such as debit, ATM, stored value, gift, fleet, and Electronic Benefits Transfer (EBT). Chip and PIN (EMV) transaction flow is similar to magnetic stripe cards, but EMV cards have an enhanced cardholder authentication mechanism based on cryptography thanks to a built-in chip.

Plastic cards, both magnetic stripe and EMV, in a sense, are virtual digital money, but they are not a digital currency because they just provide an easier method of managing payments using the same traditional fiat currencies. However, there are multiple privacy and security issues and concerns associated with plastic cards.[4] Therefore, the payment industry has long been in search of new, alternative methods of payments. It is possible that plastic payment card technology will be eventually adapted in some way to carry cryptocurrency and process crypto payments.[5] But in any case, such a transformation would only preserve a facade that millions of consumers around the world find very convenient. Inside, plastic card processing is quite different from crypto payments, which might completely change the industry. Even after a very brief look at basic transaction flows (shown in Figures 1-1 and 1-2), without going deeply into the details, it is obvious that bitcoin payment processing requires less participants and therefore is less expensive.

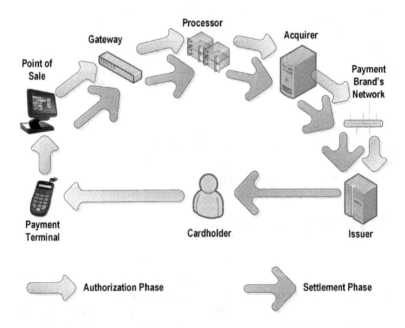

Figure 1-1: Typical Credit Card Transaction Processing Flow

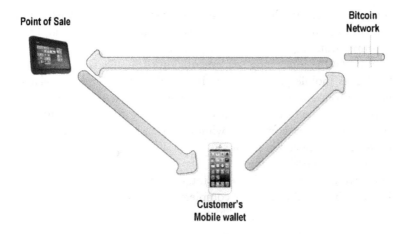

Figure 1-2: Basic Bitcoin Transaction Processing Flow

In a bitcoin transaction, there is no card issuing or acquiring bank and no payment processor or gateway.

Note: In fact, a payment processor is required in many cases in order to process bitcoin transactions between merchants and customers, especially in brick-and-mortar environments. However, a simple transaction between two individuals, for example, can be done using basic tools, without any intermediary, which is impossible in the case of credit card payments.

Nevertheless, credit and debit card payments still dominate the market of both offline (initiated in brick-and-mortar stores) and online electronic payments (although Internet transactions with credit cards are often processed through special *online payment processors*, which are discussed later in this chapter).

Mobile Payments

Even though mobile checkout is a very popular and promising trend, in most cases it is no more than just another extension of traditional fiat money that uses either the banking system or credit card infrastructure or both. Mobile payments usually introduce a new way of interaction between the customer and the point of sale (POS), with the same information being entered into the payment system (such as a token representing the banking account or credit card number). There are different technologies currently used to exchange information between the mobile wallet and the POS: Near Field Communication (NFC), barcode scanners, and even magnetic field emitters.[6] One interesting implementation of mobile checkout is the Starbucks mobile payment app.[7] It uses design principles similar to what I proposed in my 2009 white paper, "Mobile Checkout".[8] The app displays the QR code, which is scanned by the POS scanner, so the transaction is completed without physical contact between the customer device and point of sale.

Bitcoin mobile wallets also use QR codes to exchange data between the wallet and the point of sale, but the process that happens behind the scenes is completely different.

From Coins to Crypto

The chronology of world history shows that time is running much faster these days. If in the past there were hundreds or thousands of years between major inventions, nowadays major developments happen in a matter of a few years or even months. It's not just intuitive assumption—in fact, there is a scientific theory and mathematical model that confirms these observations.[9] Physicist and demographer Sergei Kapitza explains why historical periods eventually become shorter and shorter.[10]

At the beginning of humanity in the Lower Paleolithic, the interval for substantial change was roughly a million years. During the Middle Ages the period of change was a thousand years, and currently the period of change is only 45 years.

He says that the development of mankind seems to have sped up a thousand times from the Lower Paleolithic to the Middle Ages. This phenomenon is well known to historians and philosophers. Historical periodization should not be measured by astronomical time but rather by the proper time of the system. Such proper time is determined by the population growth: the more people live in the world, the higher the complexity of our system, and the faster it flows. If we assume that history is measured by the summary of human lifetimes rather than by the number of Earth's revolutions around the Sun, the shortening of historical periods gets an instant explanation. At the beginning of the Paleolithic period, the population of our ancestors was only about a hundred thousand, so the total number of people who have been living during the

Paleolithic period was about 10 billion. Exactly the same number of people has passed through the Earth for a thousand years of the Middle Ages and for 125 years of recent history. Nowadays, 10 billion people live on Earth during just a half-century. The entire historical era has shrunk to just a single generation.

Figure 1-3: Time Intervals between Major Inventions in Payment Systems

Figure 1-3 shows the shortening time intervals between major events related to the development of payment systems. The first coins were created in Lydia (now Turkey) in sixth century BC. The ancient Greeks were the first society that accepted new inventions. In the Western world, it took more than a millennium until the creation of paper money in the seventeenth century. The first bank checks, however, were already in use by Italian banks in the fourteenth century.[11] The automatic clearing houses started working in the 1970s. Diners Club created the first credit card in 1950. The first Internet payment system using "gold money"—e-gold—was introduced 46 years later in 1996, and then payment systems with the independent digital currency started right after it in 1999. And finally, the first crypto payment using the bitcoin network was performed in 2009, just 13 years after implementation of first online currency. We can see how time intervals between those major events in history of payments have been reduced from a magnitude of thousands to just a few years.

The first bank in the world was created by the Knights Templar, and it was active for almost 200 years until the full collapse of the Order of the Temple in 1314. But the failure of the first bank wasn't the end of the banking system, which flourishes to this day. The first and once biggest bitcoin online exchange service, Mt. Gox, collapsed in 2014, just four years after it was launched in 2010. But the fact that its lifespan was so short does not necessarily mean that the entire bitcoin system is broken. It's just that time runs faster according to the explosive growth theory.

Since creation of metal coins more than two and a half millenniums ago, new types of payment tokens and methods have not displaced their predecessors, but instead they just extend the assortment of possible *tenders* that can be accepted at the register. We can pay at the retail store using credit and debit cards, PayPal, and in some cases with bitcoin, but metal coins, paper banknotes, or bank checks still are very welcomed by merchants. Compare this with other technologies: once we started downloading movies online, we almost immediately forgot about DVDs (and video cassettes, respectively). The same thing happened with CDs, cassettes, and LPs. As soon as we discover more convenient technology, we quickly forget about the old one. This is not the case with payments for several reasons. First, there is no single payment technology that would fit all the possible situations: online, brick-and-mortar, remote money transfers, face-to-face payments, and so on. Second, merchants want to please as many groups of buyers as they can. There are people who have no clue about bitcoins. And there are people who don't have access to a banking system (credit cards, checks, etc.).

Merchants just want everyone to be able to buy their goods. And finally, there is another strong reason for keeping old payment methods: anonymity. Transition from coins and paper money to bank checks and electronic payments introduced a privacy issue: the owner of a bank checking account or a credit card can be identified and traced, so transactions could not be anonymous anymore. Even though cash might seem to be an anonymous payment method at first glance, this isn't exactly true: banknotes can be marked by invisible ink or radio isotopes, their unique serial numbers can be simply written down, and they can be scanned for fingerprints or DNA. Thus cash, in fact, is *pseudonymous*—just like bitcoin.

CHAPTER 2

Digital Gold

The golden rule is that there are no golden rules.
—George Bernard Shaw

There is a science fiction novel by Alexei Tolstoy[12] called *The Garin Death Ray*, also known as *The Hyperboloid of Engineer Garin*.[13] In his novel the author tells the story of a Russian engineer who creates a "hyperboloid"—the device that emits a heat ray of tremendous power, capable of destroying any obstacles. The device got its name because of the design, which consisted of two hyperbolical mirrors made from astronomical bronze and *shamonit* (fictional carbonic mineral with high refractoriness). Hyperboloid, in a sense, is a prototype, or prediction of the invention of modern lasers, although the novel was written 30 years before the first real "death ray" was created in 1957.[14] Garin escapes Soviet Russia for France and finds funding for manufacturing various types and sizes of his hyperboloids. Eventually, he captures an uninhabited island in the Pacific, where, using his hyperboloids, he starts excessively mining gold from previously inaccessible depths of the Earth. With access to virtually unlimited supplies of gold, he undermines the international gold prices, which puts the world into a severe financial crisis. As a result, Garin buys the entire US industry and becomes a world dictator.

The novel was written in 1927, long before the United States stopped tying the dollar to gold in 1971. Back then, gold meant money, and money meant gold. Nowadays, gold still equals money, but not vice versa. State currencies such as the dollar and the euro became *fiat money*—money that is regulated by the state.

Gold Standard

All money is created in the form of credit (new debt). If all loans were to be paid off, all money would disappear. Because interest has to be paid on every loan, however, more and more new money (i.e., debt) has to be created. We call money that is created during this process of unbacked money creation *fiat*, or fiduciary, money. Its value rests on the confidence that goods or services can be paid for (the term fiat is derived from *fiat lux* in Latin, which means "Let there be light."[15])

The fact that traditional money is not guaranteed by any state or by private gold reserves anymore in a sense made the creation of virtual digital currencies possible: if an official fiat currency is not backed by gold, why can't an alternative private currency exist?

Nowadays a sudden excessive supply of gold probably would disrupt the world economy, but it would unlikely crash it completely. Ironically or naturally, however, the first truly digital currencies (Table 2-1) were tied directly to gold, just like their shiny real-life predecessors in the glorious past.

Table 2-1: Defunct Digital Gold Currencies

NAME	BORN	DIED	USERS	CURRENCY
e-gold	1996	2008	2.500,000	1 g of gold
e-Bullion	2001	2008	1,000,000	e-currency
1mdc	2001	2007	N/A	e-gold
Pecunix	2002	N/A	N/A	GAU – Gram of Aurum (*lat.* Gold)

E-Gold

E-gold was founded in 1996 by Gold & Silver Reserve, Inc. The system was operating with precious metals—silver, gold, platinum, and palladium—which were stored in Gold & Silver Reserve vaults. The users could create accounts by buying one of the precious metals at its market price. This feature made e-gold very convenient and popular for processing international payments because user accounts were not tied to any national currency. In addition, the fluctuating rates of fiat currencies had no effect on e-gold account owners (unlike the fluctuations of the market price of precious metals).

In May 2007 a federal grand jury indicted e-gold, accusing the company of money laundering, conspiracy, and operating an unlicensed money-transmitting business.

Federal authorities accused e-gold of helping criminals collect and transfer millions of dollars in ill-gotten gains. The stakes were high, the government alleges. Criminals operating an investment scam using 10 specific e-gold accounts were able to move $146 million through the e-gold system, the indictment says. It also accuses the company of knowingly allowing child pornographers to move money through the e-gold system.[16]

It was the beginning of the end of the first successful digital currency and payment system. The company shut down its operations completely in 2008.

E-gold is probably the most notable digital gold currency because it was the first digital payment system that introduced, back in April 1998, one of the most important features of digital payment systems: the application programming interface (API) which allowed merchants to accept e-gold payments.[17] The API specification, called the "e-gold shopping cart interface," defined the interaction between the three parties of online payment process using digital money: the buyer (the payer, merchant); the payee; and e-gold, the payment processor (as shown in Figure 2-1).

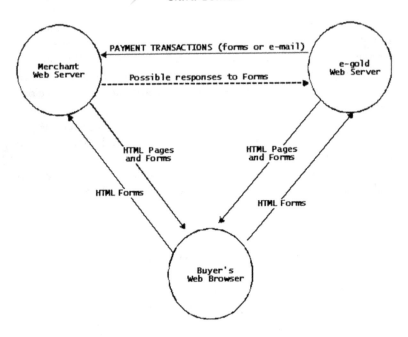

Figure 2-1: e-gold Online Payment Transaction Flow
Source: e-gold Shopping Cart Interface specification

The API specification included description of flows and messages of payment, void, and return transactions, as well as security features intended to protect confidentiality and integrity of the messages between the customer's browser, the merchant's website, and the e-gold payment processing server (as shown in Figure 2-2).

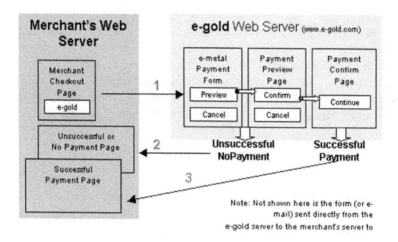

Figure 2-2: e-gold Online Payment Transaction Flow
Source: e-gold Shopping Cart Interface specification

The confidentiality of the messages was provided using the Secure Socket Layer (SSL) protocol, while the integrity was provided using the MD5 hash function (a simple analog of the digital signature). The disadvantage of the MD5 hash when compared to a full digital signature is that in addition to integrity the latter also provides the authenticity protection (you can find more detailed information about hash functions and digital signatures in chapter 5).

E-gold also was probably the first online payment service that was operating outside of the traditional financial and payment systems, meaning it had introduced both key components of real digital money: *medium of* exchange (e-gold units equivalent to a particular quantity of gold) and an *exchange system* (online payment service, merchants' API, and integration with an online shopping card). However, it wasn't really a genuine independent money machine because it was *centralized digital currency*, which is linked to gold reserves stored in bank deposits. Bitcoin became the first *decentralized digital currency* to overcome this barrier and become a digital currency that was fully abstracted from the traditional financial systems.

Now try to replace the word "e-gold" with "US Dollar," "Bank of America," "PayPal," or "bitcoin." The indictment statement still sounds sensible. However, there is a difference between bitcoin and gold currencies such as e-gold. It was relatively easy to shut down e-gold and Liberty Reserve because they both were privately held companies with centralized systems. Such systems can be destroyed by arresting their owners and terminating their servers. National currency can be destroyed by country occupation and declaration of new currency. Bitcoin can't be destroyed by the same means simply because it does not have owners or a specific central location.

e-Bullion

e-Bullion, another digital gold payment system, was founded by Jim Fayed and his wife, Pam Fayed, in 2001. A user could choose from three types of funds storage in an e-Bullion account: weight units of gold or silver, or US dollars (e-Bullion Gold, e-Bullion Silver, and e-Bullion Currency respectively). If a user chose one of the first two options, the account balance constantly changed depending on the market value of gold or silver.

E-Bullion did not charge any fees for deposits or money transfers between internal accounts. However, all these advantages were "compensated" by a monthly, fixed "service fee" for account maintenance. In addition, e-Bullion had a "liquidation fee" for account cancellation which was 3 percent for e-Bullion Gold and Silver and 2 percent for e-Bullion Currency. The truth was that it was just the fee for the withdrawal of funds from the account. This fee was charged for any withdrawal by bank transfer or check.

One of the most interesting features that distinguished e-Bullion from other payment systems was the fact that they probably were the first to use *two-factor authentication* in order to protect user accounts. Of course, this security measure was an option due to its cost (about $100). Users who were interested in higher security and were ready to pay the fee could order a special CRYPTOCard, which was made using CRYPTOCard Secure Password Technology. This card, which was implemented by e-Bullion in 2002, could be sent by mail to any place in the world. With CRYPTOCard, users could log into their accounts from any computer without having to worry about the safety of the account, whereas such operation was not recommended with regular accounts or other systems.

CRYPTOCard Secure Password Technology replaced static passwords by combining two-factor authentication with one-time passwords (OTP) to prevent the use of lost, stolen, easily guessed, or shared passwords to gain access to protected systems. Once implemented, only users to whom you have issued a CRYPTOCard token would be able to gain access to the protected network, systems, or resources. With each login attempt their token will provide a new and unique password, valid only for the specific user and the current login attempt.[18]

CHAPTER 3

Centralized Digital Payments

Everyone has as much right as he has might.

—Benedict Spinoza

Since we are going to deal with crypto payments, we should understand the difference between two terms that are both associated with money: *currency* and a *payment system*. Digital currency is an independent medium of exchange which has a value and can be exchanged for goods, services, or other currency. A payment system supports the currency (including digital currency) by providing a method of exchange between different people and entities. The dollar and euro are currencies, while cash and credit cards are payment systems.

Felix Martin, in his book called *Money: The Unauthorized Biography*, provides the following definition of money:

Money is not a commodity medium of exchange, but a social technology composed of three fundamental elements. The first is an abstract unit of value in which money is denominated. The second is a system of accounts, which keeps track of the individuals' or the institutions' credit or debit balances as they engage in trade with one another. The third is the possibility that the original creditor in a relationship can transfer their debtor's obligation to a third party in settlement of some unrelated debt.[19]

Note that cryptocurrencies such as bitcoin perfectly fit the first two elements of this definition. But let's take a look at the third element, which can be translated from financial to technical language as a payment system. In addition to several digital gold currencies, there are multiple (over 200[20]) e-payment systems that are not tied to gold deposits and are divided into two groups: digital currencies and online payment processors (as listed in Table 3-1 and Table 3-2 respectively).

The first, smaller group tried to introduce its own currency as a medium of exchange for Internet payments, while the second, larger group just facilitated online payments using traditional fiat currencies as media of exchange. Another notable difference between the two is that most attempts to create centralized digital currencies ended as fiascos (Liberty Reserve, e-gold), while multiple online payment processors currently flourish (PayPal, Amazon Payments).

The dollar is also a kind of digital currency in a sense because only 8 percent of all the dollars in the world exist as paper or coins, while the vast majority of most acceptable fiat money is stored as digital records in bank computers. However, despite such a close connection, the Internet environment is not a native habitat of the dollar. There are various multilayer electronic superstructures built around the dollar that allows it to be kept in virtual form.

Table 3-1: Defunct Centralized Digital Currencies

NAME	BORN	DIED	NUMBER OF USERS	CURRENCY
Flooz	1999	2001	150,000	flooz
Beenz	1999	2001	Unknown	beenz
Internet Cash	1999	2001	Unknown	e-currency
Liberty Reserve	2007	2013	1,000,000	"LR Dollar" or "LR Euro," linked to US dollar or euro respectively

DigiCash and ecash

Although credit cards still remain the most popular method of electronic payment since they first became popular several decades ago, their security has significant flaws embedded in the original design that took place before the beginning of the Internet era. This fact encouraged many enthusiasts who tried to introduce privacy and security to online payments.

In 1983 David Chaum proposed the use of cryptography for implementing digital payments.[21] He founded a company called *DigiCash*, which provided privacy and security to payment transactions using his crypto invention, *blind digital signatures*.[22] In 1993 Chaum invented the digital payment system *ecash*. The ecash payer was anonymous but "under exceptional circumstances" could reveal her or his identity, for example, in order to provide proof of payment.[23, 24] The anonymity of the payer was achieved by blinded signatures using the RSA (Rivest, Shamir, Adleman) encryption algorithm.

According to insiders, it was a technically perfect product that made it possible to safely and anonymously pay over the Internet. This was a field in which a lot of work needed to be done, according to the ever-paranoid cryptographers. They considered paying with a credit card to be extremely insecure. Someone only had to intercept the number to be able to spend someone else's money. Credit cards are also very cumbersome for small payments. The transaction fees are simply too high. Ecash, however, was perfectly suited for sending electronic pennies, nickels, and dimes over the Internet.[25]

Even though ecash already contained elements of modern cryptocurrency, there was a fundamental difference: it was dependent on central management by a financial institution, which processed the actual financial transactions by debiting the payer account and crediting the payee account. The entire payment process was supposed to be implemented and orchestrated by banks. Therefore, in order to pay or get paid with ecash, both the consumer (the payer) and the merchant (the payee) had to open bank accounts, as shown in the diagram in Figure 3-1.

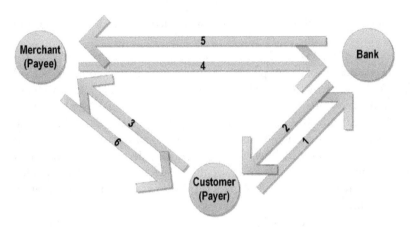

Figure 3-1: ecash Payment Flow

Ecash software, when it was first installed and executed on a payer's computer, generated a pair of private and public keys. The private key was kept secret and used to sign the ecash transactions originated from the payer (isn't it similar to bitcoin?). The public key was available for banks, merchants, and anyone else who wanted to verify the message and ecash transactions initiated by the payer (bitcoin, once again!). When the payer needed to send a payment, the following steps were performed:

1. The ecash software generated the "notes" for each required denomination according to the total amount of the transaction. A set of standard fixed denominations protected the identity of the payer by using the unique transaction amount. In addition, the payer's client software generated a random "blinding" factor that was used to blind the denominations in order to provide anonymity (such as hiding the exact total transaction amount to make the payer impossible to identify from a relatively unique number such as $1,327.89). The blinded denominations were then encrypted with the bank's public key and sent to the bank.

2. The bank decrypted the message using its private key, signed the blinded notes with its private key, debit the payer account, and sent the signed blinded denominations back to the payer.

3. The payer received the signed notes from the bank—and here is the most important step—took out the blinding factor. The

bank signatures still remained valid, and the bank could now validate those signatures but was unable to identify the payer. So the original notes, along with the bank signatures, became the actual ecash, and the payer could send them to the payee.

4. Upon receipt of the ecash (original notes plus bank signatures), the payee validated it using the payer and the bank public key, and then could sent the message to the bank for further verification, double-spending check, and actual payment processing.

5. The bank confirmed the validity of the ecash, checked if it wasn't spent already, and credited the payee account.

6. The payee then sent the payment confirmation back to the payer.

Perhaps the most important feature of ecash was support of both online and offline transactions. The payee could accept ecash payment offline because it was possible to validate the payer and the bank signatures offline. Of course, the offline transaction was not protected against double-spending; however, the bank could recognize the fraudulent transaction after the fact and identify the payer. Bitcoin offers a different, more elegant, completely decentralized, but not less powerful, anti-double-spending mechanism.

Online Currencies: Flooz and Beenz

Beenz was one of the attempts to create a "native" Internet currency. It was launched in 1998 by Charles Cohen who said, "I believe we'll start to see beenz listed against other major currencies." Here's how beenz worked according to TIME magazine:

First, you open a free account with Beenz.com, a New York City-based outfit that's been in business since 1998. You then earn beenz by visiting certain websites that give beenz away as a means of rewarding customer loyalty in exchange for personal information or as a reward to surfers for just showing up. Then, once your virtual wallet is bulging, you can spend beenz at any of the 200 e-commerce sites that accept beenz as a form of payment.[26]

The company eventually closed its operations in 2001.[27]

Flooz started in 1999 and also died in 2001. Similar to Beenz, it was trying to invent online currency but failed after it fell victim to a softening economy due to the dot-com bubble collapse as well as "a ring of credit card thieves operating out of Russia and other parts of eastern Europe...suspected of using stolen cards to buy the [flooz] currency."[28]

Liberty Reserve

Liberty Reserve was not a pure digital gold currency. It acted as an intermediary between digital gold and the payment processors (which are described later in this chapter) because the users' Liberty Reserve funds could be stored as an LR Dollar, LR Euro, or LR Gold, which were tied to the dollar, the euro, or gold, respectively. Liberty Reserve could process payments as well as money transfers as shown in Figure 3-2.

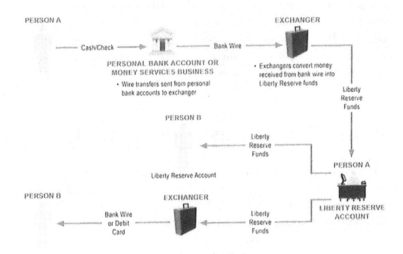

Figure 3-2: Liberty Reserve Money Transfer Scheme
Source: Department of the Treasury[29]

The user accounts were protected from hacking by several security measures. In addition to using a regular password, there was an additional protection level: the user's PIN code, which was created during registration. Instead of typing on a regular keyboard, the user dialed this PIN using a virtual keyboard (as shown in Figure 3-3), which provided protection against malware (keystroke loggers). Moreover, the location of the keys on the virtual keyboard was not permanent; it changed every time the page was loaded, which prevented hackers from using programs that can store mouse movements and clicks.

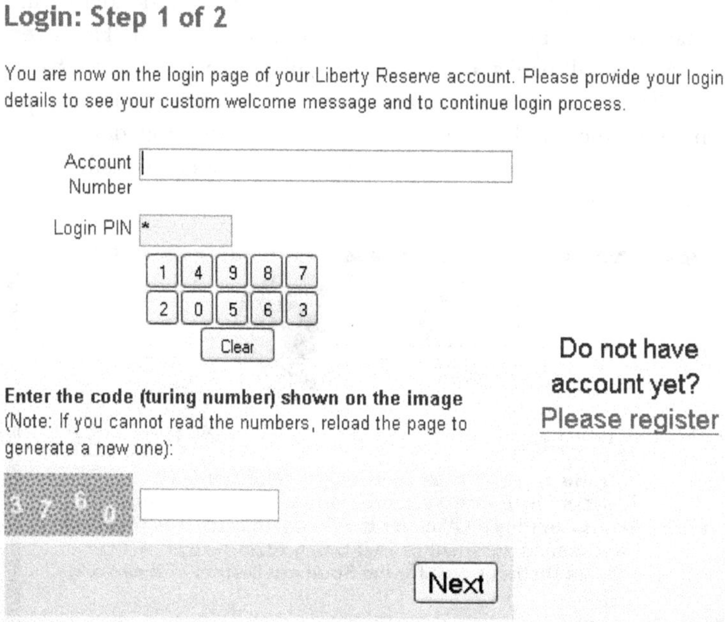

Login: Step 1 of 2

You are now on the login page of your Liberty Reserve account. Please provide your login details to see your custom welcome message and to continue login process.

Account Number

Login PIN [*]

| 1 | 4 | 9 | 8 | 7 |
| 2 | 0 | 5 | 6 | 3 |

Clear

Do not have account yet?
Please register

Enter the code (turing number) shown on the image
(Note: If you cannot read the numbers, reload the page to generate a new one):

Next

Figure 3-3: Virtual Keyboard for Entering the PIN on the Liberty Reserve Login Screen

After successfully completing the PIN entry, the system prompted the user to enter the regular password, which could be typed using either the virtual or regular keyboard. In addition, the system displayed the personal welcome message that was previously created by the user during the registration process to protect against fishing attacks. If the message displayed to the user during login is different from the one he or she chose at registration, or there is no image at all, it indicated that the login web page was faked and that the user should immediately abort the login attempt and scan the system for malware. Even today many banks still do not have such sophisticated security controls.

Nevertheless, eventually, it was found that Liberty Reserve facilitated money-laundering activities, and on May 2013 its funds were confiscated, the domain was seized, and the website was closed by law enforcement agencies (Figure 3-4). The indictment against Liberty Reserve contains interesting implementation details which emphasize the anonymous character of the system:

In registering, the user was required to provide basic identifying information, such as name, address, and date of birth. However, unlike traditional banks or legitimate online payment processors, Liberty Reserve did not require users to validate their identity information, such as by providing official identification documents or a credit card. Accounts could therefore be opened easily using fictitious or anonymous identities.[30]

Figure 3-4: Liberty Reserve and 35 Exchangers' Domains Have Been Seized
Source: Liberty Reserve website[31]

The collapse of Liberty Reserve in 2013, which was preceded by a similar failure of other centralized digital currencies such as e-gold, e-Bullion, and DigiCash, finally cleared the road for decentralized crypto payment systems such as bitcoin, exactly in the same way as the shutdown of the once popular free music download site Napster in 2001 inspired creation of peer-to-peer content sharing technologies such as BitTorrent.[32]

However, the way in which this defunct payment system has been destroyed teaches us a lesson that should be taken into account before trusting cryptocurrencies. Liberty Reserve was using third party exchangers in order to deposit and withdraw funds from user accounts, which is very similar to how the bitcoin network functions:

To add an additional layer of anonymity, Liberty Reserve did not permit users to fund their accounts by transferring money to Liberty Reserve directly, such as by issuing a credit card payment or wire transfer to Liberty Reserve. Nor could Liberty Reserve users withdraw funds from their accounts directly, such as through an ATM withdrawal. Instead, Liberty Reserve users were required to make any deposits or withdrawals through the use of third-party "exchangers," thus enabling Liberty Reserve to avoid collecting any information about its users through banking transactions or other activity that would leave a centralized financial paper trail.

Liberty Reserve's "exchangers" were third-party entities that maintained direct financial relationships with Liberty Reserve, buying and selling LR in bulk from Liberty Reserve in exchange for mainstream currency. The exchangers in turn bought and sold this LR in smaller transactions with end users in exchange for mainstream currency.

The Liberty Reserve website recommended a number of "pre-approved" exchangers. These exchangers tended to be unlicensed money-transmitting businesses operating without significant governmental oversight or regulation.

Eventually, not just the Liberty Reserve website and domain were destroyed, but most third-party exchangers—a total of 35 domains—were shut down.[33] Thus, even though the payment processor theoretically could restore itself—for example, by setting up new servers in a different country and modifying the domain name—the whole payment network was already distrusted because its users could not make new deposits or withdraw their savings.

Unlike centralized payment systems, bitcoin and other cryptocurrencies cannot be shut down just by targeting a single domain. But their exchangers are still based on the same principles as their predecessors as they are centralized, privately owned companies located in a particular jurisdiction. By destroying local exchangers, national governments can distrust a particular crypto payment network even though they are unable to destroy the cryptocurrency as a whole.

Online Payment Processors

Nowadays, PayPal, Amazon Payments, WebMoney, and many other *online payment processors* dominate the market of Internet payments (major processors are listed in Table 3-2). In a nutshell, online payment processors are simply extensions, or superstructures, of the banking and credit card systems, which after all, use fiat money as a medium of exchange. Therefore, they are a great invention of their time; however, they cannot be classified as digital money.

The major achievement of online payment processors is that they facilitate Internet transactions using traditional bank accounts and plastic cards by making them more secure and convenient for both buyers and sellers. They usually provide a sophisticated API so developers can seamlessly integrate payments into their products. For example, Amazon Payments, which is less popular beyond the Amazon website than its rival PayPal, provides an API function, among other things, which allows developers to facilitate payments between buyers and sellers.

Table 3-2: Online Payment Processors

NAME	BORN	DIED	USERS	CURRENCY
DigiCash	1990	1998	Unknown	Denominations of national currencies
WebMoney	1998	Active	25 million	WM unit linked to various national currencies or bitcoin
PayPal	2000	Active	148 million	National currencies

| Amazon Payments | 2007 | Active | Unknown | National currencies |

The major disadvantage of online payment processors, besides the fact that they are still dependent on the traditional banking and payment card industry, is that they operate mainly online and have little presence in brick-and-mortar stores. However, lately, this trend is also changing as online payment processors attempt to enter the mainstream with their new payment tools.[34]

CHAPTER 4

Cryptocurrencies

Freedom, Security, Convenience: Choose Two.[35]

—Dan Geer

There are many questions around the future of bitcoin. Will it eventually be accepted as a mainstream method of payment? Or, putting it in a more dramatic way, will it exist at all? Let's face it: just a few years have passed since the invention of bitcoin, and despite many thousands of users paying with it, many merchants accepting it as legitimate tender, and various companies servicing both those users and merchants, this question is still open.

If the majority of national governments decide that they don't want to see the powerful and unregulated alternative to their national fiat currencies and traditional payment systems anymore, together they are still able to doom it. However, what they can't do is destroy the principles of bitcoin. Those principles are good and right, and therefore they will probably stay with us for a long time, even if they are implemented in a different form using different, perhaps even better, technologies. They were once formulated in the Satoshi Nakamoto white paper,[36] and then enhanced and developed by the bitcoin community. Even today, there are more than 200 cryptocurrencies that implement bitcoin principles, and it's hard to believe that they will disappear.

Satoshi Nakamoto White Paper

The principles of bitcoin design were proposed in a technical white paper published by Satoshi Nakamoto in 2008. To date nobody knows who is hiding behind this name—whether it is a real person, a pseudonym, or a group of individuals. There were several attempts to identify the author(s), but nothing was concluded with realistic results.[37, 38, 39, 40]

Nevertheless, in one of the letters explaining the white paper, Satoshi Nakamoto defines the main properties of bitcoin:

- Double-spending is prevented with a peer-to-peer network.
- There is no mint or other trusted parties.
- Participants can be anonymous.[41]

Let's try to break down these points into several concepts so we can understand them.

Double-Spending Problem

Double-spending is one of the main problems for electronic payments in general. In the plastic cards world this problem is resolved (partially) by a single point of authorization. All credit and debit transactions eventually come to the issuing bank for authorization. If the cardholder's account has enough credit (or in case of debit cards—enough funds) available to cover the transaction amount, the issuer approves the payment.

Why is this problem partially resolved? Because if the authorizer's system is not available (network is down), many merchants will use a special function that is known under several different names with the same meaning: stand-in, fallback, offline approval, or store and forward. It means that someone or something—it could be a point-of-sale application, store payment server, payment gateway, or payment processor—may automatically authorize the transaction without actual communication with the card issuer. Since only the card issuer knows the real credit limit or balance on the cardholder's account, such transactions are always vulnerable to double-spending. If I know that the payment system at store 1 is working in offline (fallback) mode, I can spend a whole amount on my card simultaneously at stores 1 and 2, and I can spend twice the actual amount available on my account (Figure 4-1).

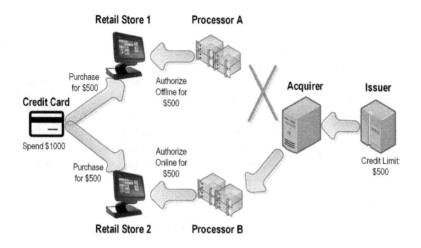

Figure 4-1: Double-Spending

In theory, overspending can equate to more than double-spending if you find another store 3 that is also in fallback mode. In many situations, the payment system goes automatically to offline mode, for example, due to communication failure. However, there are scenarios when payment systems wittingly switch to offline approval mode—for example, in order to save on communication or transaction fees when the transaction amount (read *risk of money loss*) is relatively small. In any case, the merchant is the one who is taking the risk and pays in case of chargeback—the situation that occurs when the issuer rejects paying for a transaction when it comes to the settlement phase and finds that the cardholder account is empty.

At first glance, in a world of online cryptocurrencies like bitcoin, the double-spending problem may seem unsolvable because there is no central entity—such as credit-card-issuing bank—that would be solely responsible for a decision to approve or reject the payment. However, a special mechanism was designed in order to prevent double-spending. This mechanism is based on several other concepts: *peer-to-peer networking, proof of work,* and blockchain. Since there are many details associated with each of these, and they are all linked to each other, we are going to review them in chapter 4. At this point you can just understand that a built-in double-spending prevention mechanism is one of the most powerful features of bitcoin design.

Decentralization

One may ask a simple question: if the fact that bitcoin is both currency and a payment system prevents it from gaining mainstream acceptance, wouldn't it be easier to take just the crypto payment element and link it to the existing fiat currency? Wouldn't the resulting system—let's call it "cryptodollar"—be a great replacement for insecure credit and debit networks? I do not rule out the possibility that there are big corporations and small startups already working in this direction. But the problem is the maintenance of such a system. Bitcoin is not maintained by just enthusiasts, at least not today. Those people earn money not just by transaction processing but also by *mining*—a process of creating new bitcoins. The same incentive can't be used directly in the cryptodollar system because the Federal Reserve has exclusive rights to manufacture the new dollars. Therefore, all the cryptodollar processing nodes would need to be maintained by (or associated with) the Federal Reserve, which contradicts one of the main principles (and advantages!) of cryptocurrency—decentralization.

As we discussed in chapter 1, the reason for the failure of some pre-bitcoin virtual currencies was their *centralized* management. There was always a single site, or well-defined group of sites, that was responsible for maintaining the availability and *integrity* of the payment system. *Integrity* here means that there should always be a *trusted party*, such as a website in case of virtual currency, or an issuing bank in case of credit card payments, that is responsible for maintaining the user account balance.

Privacy: Anonymity or Pseudonymity?

Availability is one of three key principles of information security (in addition to *confidentiality* and *integrity*) which is often overlooked, or sometimes even considered a nonsecurity topic. However, when it comes to existence of the system, whether it is a social network or virtual currency, availability is the key feature that can be targeted using denial of service (DoS) attacks, for example. In case of online currencies, such as Liberty Reserve, the law enforcement agency simply obtained an order to shut down all the relevant websites, which effectively killed the entire system. In order to be resistant to DoS or government attacks, bitcoin employs a *peer-to-peer network*, which is "based on cryptographic proof instead of trust, allowing any two willing parties to transact directly with each other without the need for a trusted third party." Such an approach gives us hope that bitcoin will be able to sustain most technical and political challenges that could emerge as cryptocurrencies become more popular and start seriously competing with big corporations and national governments.

In today's world of total electronic surveillance, it's difficult to be an active player in financial life while enjoying complete privacy, regardless of which method of payment you choose. Unfortunately, bitcoin technology is not an exception.

Satoshi Nakamoto's white paper contains the following definition of bitcoin's "new privacy model," comparing it to the traditional approach used by the financial industry (Figure 4-2):

The traditional banking model achieves a level of privacy by limiting access to information to the parties involved and the trusted third party. The need to announce all transactions publicly precludes this method, but privacy can still be maintained.

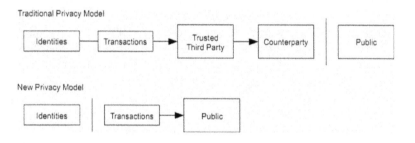

Figure 4-2: Bitcoin Privacy Model
Source: Satoshi Nakamoto white paper

Unlike the traditional approach used by banks, where financial transaction records are kept secret, along with the identities of their owners, bitcoin transactions are stored in a public registry, or *ledger*, called *blockchain*, which is freely accessible to anyone, anytime. In this situation, the user's identity is kept private due to the anonymity of bitcoin accounts, which are not linked to particular individuals. When creating the bitcoin wallet, there is no need to provide and prove your identity—something that is mandatory when you open a bank checking or credit account. There are advantages and disadvantages to this approach. On the one hand, there is no one who knows (and therefore can potentially disclose) your identity. On the other hand, all your transactions are permanently visible to anyone, and therefore can be potentially traced to your identity using high-tech methods of surveillance and analytics.

In chapter 3 we briefly discussed the ecash technology, which was invented and developed in the 1990s by David Chaum, who once said, "The difference between a bad electronic cash system and well-developed digital cash will determine whether we will have a dictatorship or a real democracy."[42] Perhaps ecash was the first serious attempt to resolve the privacy issue associated with electronic payments. Similar to ecash, bitcoin does not really provide full privacy and anonymity, but rather *pseudonymity*. There are at least two reasons.

First, bitcoin transactions are fully traceable because they are stored in blockchains, which can be downloaded and analyzed by anyone who has Internet access. Bitcoin transactions are still anonymous as there is no direct link between the payer identity and the bitcoin wallet—anyone can create any number of bitcoin addresses, which are simply a series of bytes, and use them anonymously. However, the very fact that transaction records are not secret makes anonymity incomplete and vulnerable because theoretically it is still possible to link the identity to the address, for example, by tracing the IP address of the machine where the wallet resides. Once the ownership of the address is determined, all the past and future payments associated with this wallet can be traced to its owner by analyzing the blockchain.

Another vulnerability of bitcoin, which makes it a pseudonymous rather than an anonymous system, is the way the funds are deposited into or withdrawn from the bitcoin network. We still live in the real world, which is far away from total adoption of cryptocurrencies. Therefore, we need a way to trade our day-to-day fiat money for bitcoins in order to make a payment, or exchange bitcoins for bucks or other national currency to accept the payment. Unlike many other electronic payment systems, the bitcoin network itself is isolated from the exchange services that trade fiat currency for bitcoins and vice versa. This isolation is good and bad. It's good because by looking at only bitcoin blockchains it is impossible to tell where the money originally came from. It's bad because we rely on third-party exchange services, which can be compromised just like any other financial institution, and spit out their trade records.

Unlike traditional money, which is controlled and managed by national governments and commercial banks, cryptocurrencies are governed by the rules defined by computer programmers. This is one of the major differences between the traditional and cryptocurrencies that makes bitcoin extremely transparent, understandable, and therefore trustworthy.

Blockchain

Although bitcoin implementation consists of multiple elements, blockchain without a doubt is both the heart and the brain of any cryptocurrency. In order to understand how it works, let's try to imagine giant virtual necklace made of transaction records instead of stones or pearls. Each bead, which is called a block, consists of multiple transaction records (the number varies from a few hundred to more than a thousand[43]). In beadwork, in order to create a real piece of art, you cannot just randomly add different types of beads. Instead, you should follow some patterns and rules, so before you can string another bead you need to make sure it matches the color, form, and other features of the previous bead, usually in a way only known to the artist. In a virtual world of cryptocurrencies, the new block of transactions also cannot be connected to the previous block without implementing a special procedure called proof of work (please see more details about proof of work later in this chapter in the Mining section).

The integrity and authenticity of each bitcoin transaction in each block in the blockchain is protected by a digital signature created by the owner of the bitcoin address. The difficult part about blockchains is that unlike real-world beadwork, where usually only a single artist is crafting with any particular beads, there are multiple nodes (owned by bitcoin users) trying to add blocks to the same blockchain simultaneously. Each node is a piece of bitcoin software running on a computer connected to the Internet. A copy of an entire blockchain is usually stored on each and every node. Therefore, each bitcoin user has instant access to the full report of every crypto transactions ever done since bitcoin was created.

The first node that manages to add a new block gets a bonus in the form of a newly created bitcoin—that's how mining works. As I mentioned before, there are multiple nodes trying to mine—or earn new bitcoins—at the same time. According to the rules, which are hardcoded into the bitcoin software running on each node, the node must accept the newly compiled block from another node that managed to do it faster, and propagate this change to other connected peers. But what if two nodes are disconnected (which is a very normal situation in the Internet) and created two different new blocks at the same time? Or even worse—a malicious node adds a block containing a fraudulent transaction? Then we get a fork which will be resolved once the connectivity is restored. Eventually, all the nodes will accept the longest fork, or in other words, the version of the blockchain supported by majority of the nodes. This mechanism is an answer to a phenomenon called the Byzantine Generals' Problem.

Byzantine Generals' Problem

The Byzantine army is besieging the city. It is the evening before the decisive offensive. But the Byzantine army is heterogeneous and consists of multiple separate divisions, each of which is under the command of its own general. The forces of the Byzantines is just slightly more superior than the enemy forces, so in order to win, all the generals need to agree on the exact time and place of the attack. The problem is that communication between the generals is unreliable, and messages can be either lost or counterfeited. But the communication problem is not the worst issue. Some generals are bribed by the enemy and traitors. They want to thwart the offensive by sending false messages to other generals.

In such a difficult situation, the city will be defeated only if a majority of the generals agree on the plan of the attack by relying on a majority of the messages that match, and ignoring a minority of the messages that are different (counterfeited). Bitcoin works on the same principle: the majority always wins. Therefore, the "traitors," who try to add new blocks containing fraudulent transactions to the blockchain, will be automatically detected and eventually ignored.

Mining

Now back to the miners. As I said, many nodes would like to be first in a race for the new block, and thus receive the cherished portion of the shiny new coins. But who is going to be the first to "guess" a new block? The answer is simple: the one who has more power (pretty common rule, isn't it?) In case of bitcoin it requires more processing power in order to provide proof of work by finding the appropriate hash value.

Each block has a header. One of the block header fields contains the value of the hash calculated for the previous block (double SHA-256). This way separate blocks are connected into a single blockchain. For those familiar with programming, this is what the header looks like in the code:

```
class CBlockHeader
    int nVersion;
    uint256 hashPrevBlock;
    uint256 hashMerkleRoot;
    unsigned int nTime;
    unsigned int nBits;
    unsigned int nNonce;
```

Computers and Internet connections are fast these days, so bitcoin nodes can generate new blocks almost instantly as soon as new transactions come in. In order to reduce the number of forks containing the wrong blocks, bitcoin artificially increases the time interval between adding new blocks by complicating the process. The hash value of the block must meet a special condition which is called proof of work. When translated to a binary number, the value of the hash must be less than a special global bitcoin constant called target. The lower the target, the more difficult it is to generate a block. This an example of the bitcoin target value:

0000000000000000171A8B0000000000000000000000000000000000000000000000000000

The block header has a special variable called nonce. By changing (incrementing) the nonce and recalculating the hash, the miner can eventually find a hash value that is less than the target value, which means that this new block can be added to the blockchain and accepted by other nodes.

The target, which is often referred to as difficulty, is set so that the hash can be found on average within 10 minutes (this is why at least 10 minutes is usually required for first confirmation of the bitcoin transaction). Because more and more people would like to make money using this exotic method, more and more computational power is being added to the network. In order to keep the 10 minutes interval, the software resets the target every 2016 blocks (every two weeks). In practice it means that over time finding a new block becomes a harder task. At the beginning of the bitcoin era, many people made a fortune with just a regular desktop computer. But nowadays the only way to make any money by mining is by participating in one of special communities, called pool, which consist of hundreds or thousands of miners (and their farms of desktop computers and servers). Every 10 minutes one of the mining pools "wins" a new block of transactions and gets its fix of bitcoins. Every member of the winning pool then receives a share of the bonus that is proportional to the member's contribution of the computational power. But nowadays this contribution should be huge in order to get any significant amount of compensation. Even the most powerful workstation is not enough—you need a lot of dedicated processors constantly working in parallel and crunching hash values. At some point it turned out that graphic processors (GPUs) are pretty well suited for mining (in fact, the sellers of graphic cards made a lot of money on a modern gold rush). However, even graphic cards already became ineffective as they started consuming more electricity than earning bitcoins. In order to catch up with the enormous collective computational power of the bitcoin network, also known as *hashrate*, which is steadily climbing, modern miners designed special equipment, which is worth a lot of money and consumes a lot of electricity. Amateur mining has turned into a high-tech and expensive business. Was it the goal of the bitcoin creators? We don't know.

Part I Summary

Money and payments began from barter and commodities, which eventually inspired the creation of the first metal coins more than 2500 years ago. Since the creation of coins, invention of new tenders, such as paper money, bank checks, credit cards, mobile checkout, and online currencies, did not eliminate previous payment methods. Digital gold currencies such as e-gold and e-Bullion were the first successful online payment systems that were backed by real gold. Ecash was the first electronic payment system based on cryptography.

There are two major groups of electronic payment systems: centralized and decentralized. E-gold, ecash, Liberty Reserve, PayPal, and Amazon Payments are all examples of centralized payment systems. However, e-gold and Liberty Reserve have been operated independently from traditional financial institutions. Bitcoin is the first decentralized payment system that also does not depend directly on banks, clearinghouses, or credit card networks.

The stories of e-gold, e-Bullion, and Liberty Reserve demonstrate that most attempts to create an independent, centralized e-money ecosystem that would be more than just an extension of traditional banking and credit card systems, but introduce a real alternative currency and protect user privacy, eventually have been executed and failed. Unlike centralized currencies, bitcoin cannot be destroyed simply by shutting down the group of Internet domains.

The principles of the bitcoin design, which are currently used by other cryptocurrencies, were first defined in a technical white paper published by Satoshi Nakamoto, a person (or a group of people) whose real identity is still unknown.

There are a few basic points of bitcoin design:

- Provides a solution to the double-spending problem by processing the transaction through a peer-to-peer network and storing them in a publicly accessible blockchain
- Decentralizes the "account" management, which protects transactions from DoS attacks and national governments

- Gives bitcoin users limited anonymity, called pseudonymity

Blockchains, which are bitcoin's universal "registry," consist of blocks containing bitcoin transactions. Multiple copies of blockchains are stored by network nodes that participate in bitcoin transactions.

The owners of the wallets (nodes) can earn newly created bitcoins by generating a new block of transactions and adding it to the blockchain. This process is called mining. The bitcoin network uses the proof-of-work concept to resolve the double-spending problem and prevent the addition of counterfeit transactions into the blockchain.

PART II

Bitcoin Cryptography

In This Part

CHAPTER 5

Types of Encryption

*If you want to keep a secret
you must also hide it from yourself.*
—George Orwell

From the name of the classification where bitcoin belongs—cryptocurrency—it is already clear that "crypto" is the most important component of bitcoin. But what does this mean exactly? Obviously, without an answer to this question, we will hardly be able to move on to other components of cryptocurrencies.

Two of the three main areas of information security theory—confidentiality and integrity—are essential for cryptocurrency design. The third one—availability—is also consequential, though often forgotten. Availability à la bitcoin means that your money is never lost as long as you have access to your secret key—your bitcoin address. The beauty of bitcoin is that even though it is completely virtual and electronic currency, the address can be still stored on a small piece of paper. And like any conventional money, this piece of paper can be stored offline—for example, in a bank deposit box, far away from Internet connections, which means out of reach of hackers.

While maintaining confidentiality is important in order to preserve the ownership of the currency, supporting integrity is required for keeping transaction records intact. But both confidentiality and integrity intertwine when we talk about the ability to recognize the authenticity of payment transactions as well as preventing double-spending. All those existential preconditions of successful cryptocurrency design were made possible by using cryptography.

Symmetric Encryption

Cryptography, in its original definition, is the science of hiding information from the prying eyes of those who are not supposed to see the information: an enemy during a war, a business competitor, or even a jealous spouse. Cryptography is perhaps one of the oldest disciplines. First ciphers were simple, but some of them are still in use—for example, the *one-time pad*.[44] Its meaning lies in the fact that each character in the original message—which is called *plaintext* in cryptography —is *substituted* with a new unique symbol taken from a special text—the one-time pad. The characters in the one-time pad are randomly generated, so it is impossible to predict or calculate their sequence, therefore, a one-time pad is still the simplest and most secure method of encryption.

The entire sequence of special characters in the one-time pad is called a *shared secret*, or a *key*. In order to decrypt the resulting message, also called *ciphertext*, another party of the crypto conversation must have the same one-time pad, or the same key. Encryption algorithms using the same key for both encryption and decryption are called *symmetric* (Figure 5-1).

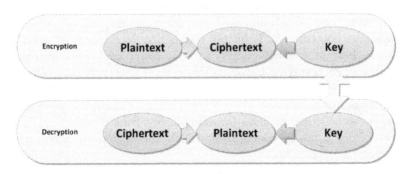

Figure 5-1: Symmetric Encryption

There are many other symmetric algorithms besides one-time pad. Some of them, such as *Caesar shift* cipher,[45] are very old and simple, and most of them are vulnerable to cryptanalysis—a part of cryptology, which deals with breaking the ciphers. Most modern symmetric encryption methods are based on complex algorithms and so they are sustainable to cryptanalysis, but only if they use a suitable key size. The size of the key usually determines the degree of the crypto system's secrecy based on the simple principle "the bigger the better." Thus, a one-time pad is the ideal cipher as its key is the same size as the size of the plaintext.

However, a one-time pad can be used only in certain situations because it must be disposable: the same one-time pad cannot be used twice to encrypt different plaintexts, otherwise the ciphertext becomes susceptible to cryptanalysis. Frequent updates required for such a bulky key are not a trivial task.

Fortunately, cryptographers came up with many other symmetric ciphers with a limited key size, which is used each time for a new fragment of the data being encoded, called a block. Therefore, those ciphers are called *block ciphers*. You may already be familiar with block ciphers, which are widely used in communication and the payment security industry. For example, the Advanced Encryption Standard (AES) is the default protocol for Transport Layer Security (TLS), also known as the Secure Socket Layer (SSL), which protects the communication between your Internet browser and the web server.[46] Another, older algorithm called the Data Encryption Standard (DES), or more precisely its upgraded version called Triple DES (TDES), is the standard method implemented in payment terminals and ATMs for debit card PIN encryption.[47]

Now hold your breath: symmetric algorithms are not used in bitcoin implementation. But you have not wasted your time reading about symmetric encryption. We had to briefly review it in order to understand how different it is from public-key cryptography, which is essential for cryptocurrency design.

The integrity and authenticity of bitcoin transactions are provided by using digital signatures, in particular the Elliptic Curve Digital Signature Algorithm (ECDSA).[48]

ECDSA is based on three main concepts:

- One-way hash functions
- Public-key (asymmetric) cryptography
- Digital signatures

One-Way Hash Functions

Cryptographic one-way hash functions are the first step towards the digital signature—a complete cryptographic solution that protects bitcoin transactions. As we can see in the diagram shown in Figure 5-

2, one-way hash functions are very useful cryptographic algorithms in bitcoin implementation. There are two hash algorithms employed by bitcoin: SHA-256 and RIPEMD-160. Let's see how they work.

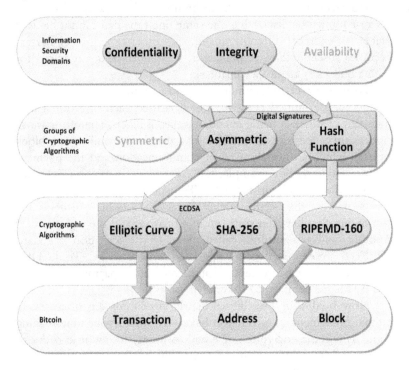

Figure 5-2: Cryptography in Bitcoin Implementation

One-Way Function and Message Digest

The idea behind a cryptographic hash function is one-way function: some operation—mathematic or otherwise—that is easy to calculate in one direction but difficult to reverse.[49] In fact, there is a special term used when describing the one-way functions: "computationally infeasible," which means that it is possible theoretically (think *brute-force* attack—a systematic loop through all possible combinations) but in practice it will take so much time that the result, once it is achieved, will not be relevant already. It is similar to the idea behind the public key, or asymmetric encryption, which we will discuss later in this chapter and further in chapters 6, 7, and 8, but there is an important difference: there is no "back door" (or "private key" in asymmetric encryption), so once the ciphertext is calculated by the hash function, there is no way to reverse it back to the original plaintext (Figure 5-3).

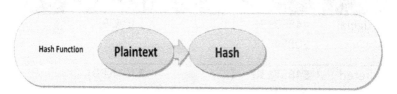

Figure 5-3: One-Way Hash Function

Another synonym for a one-way function is *message digest*, which is a good illustration of how one-way functions are useful in day-to-day life. For example, the digest formula called *mod 10* (also known as the *Luhn* formula, named after its creator, Hans Peter Luhn[50]) is used in the payment card industry in order to validate the primary account number (PAN) of the credit and debit cards.

When you take a look at the PAN, which is embossed on every plastic card, you can normally see 16 digits. But the unique part of this number, which actually identifies your account, is only nine digits. The first six digits are the bank identification number (BIN), which consists of the numbers assigned to the user by the financial institution (issuing bank). And the last digit is the *check digit*—the digest of the first 15 digits.[51]

When you swipe your card at the point of sale terminal, or especially when you or a cashier manually key in the number, the point of sale takes the first 15 digits, calculates the mod 10 digest value, and compares it with the value of the last digit it received as part of the entered PAN. If the two values match, the PAN is valid, and the POS will start processing the payment. If the result of the calculation is different from the digit that was actually entered, it indicates an error, which could be either a human mistake (when the card is manually keyed) or hardware failure (when the card is swiped). If any digit in the account number is altered, the result of the mod 10 calculation will always be different from the original check digit value. In the example shown in Table 5-1, if we change the first digit of the account from 4 to 5, it will not pass the validation procedure.

Table 5-1: Check Digit Validation using Mod 10

ACCOUNT	NUMBER	CHECK DIGIT (LAST DIGIT)	RESULT OF MOD 10 CALCULATION
Original	44857136 90854611	1	1 (PASS)
Altered	54857136 90854611	1	8 (FAILED)

Collision

The mod 10 formula is very simple and can be easily calculated with pencil and paper. It produces just 1 byte of the result, which is convenient in many applications, such as PAN check digit (which we reviewed before). However, there is a problem with mod 10 algorithms: two (in fact, many) different inputs can produce the same output. Such a feature is called *collision*. In our example with credit card numbers, multiple different accounts may have the same check digits.

A high probability of collision in the mod 10 algorithm is not an issue for the payment card industry because the purpose of the check digit is mainly recognizing the human errors, when usually only a small part of the message (one or two digits) is altered or missing. However, if the account number has been altered on purpose, there would be no way to recognize the fact of tampering if check digit has been also re-calculated. In example shown in Table 5-2, two account numbers have the same check digit 1, and so they are both valid numbers:

Table 5-2: Check Digit Validation Using Mod 10

ACCOUNT	NUMBER	CHECK DIGIT (LAST DIGIT)	RESULT OF MOD 10 CALCULATION
Original	44857136 90854611	1	1 (PASS)
Altered	51377787 57192501	1	1 (PASS)

Fortunately, there are one-way functions that can produce unique (or nearly unique) message digests that can be used to uniquely identify the input message. Those algorithms are called *cryptographic hash functions*, and they are used in many applications that require high *collision resistance*, including bitcoin. Cryptographic hash functions do not just protect the messages from occasional human or hardware errors, but provide integrity and, in conjunction with public-key encryption, authenticity of the messages (see the Digital Signatures section later in this chapter).

SHA-256

The Secure Hash Algorithm (SHA) is the cryptographic hash function used in bitcoin implementation.[52] SHA-256 produces 256 bits, or 32 bytes of output message digest. SHA-256 guarantees that collision will never occur if the function is applied to a different input message. Imagine that instead of mod 10 we use SHA-256 for card account number validations. Instead of a 1-byte check digit, this new validation mechanism would require 32 bytes of data to store the result of the hash. But in return we would get an assurance that there are no two different account numbers that would produce the same hash. All three account number values mentioned in the previous examples, as well as any other account numbers, would produce completely different SHA-256 values (Table 5-3).

Table 5-3: Account Numbers Hashed with SHA-256

ACCOUNT NUMBER	SHA-256
4485713690854611	ad51095b34842d258ccd7 d510205c63504f6d4e85b 35b57cecbda78010df7ee6
5485713690854611	77843f0d79aa28cdf482df 4d9364cf98702d91597e6 597cfc157f7197070be4e
5137778757192501	cfc282ce1716959530b40b 10d9f124c2edeef37a69b4 c8226eafdbb19e3d2f55

RIPEMD-160

While SHA-256 is used at several points throughout the bitcoin transaction lifecycle, which we will review in chapter 4, there is another cryptographic hash function employed by bitcoin—RACE Integrity Primitives Evaluation Message Digest-160 bits (*RIPEMD-160*)—which produces 160 bits (20 bytes) of output.[53] Its result is significantly smaller than the one produced by SHA-256 (Table 5-4).

Table 5-4: Output Length of SHA-256 and RIPEMD-160

ALGORITHM	OUTPUT LENGTH
SHA-256	256 bits
RIPEMD-160	160 bits

Probably the relatively compact size of RIPEMD-160 was one of the reasons it was used in bitcoin address generation.

Public-Key (Asymmetric) Cryptography

Before the public-key encryption era, the only way to secure information was by using symmetric encryption, which required users (or applications) at two ends of communication to exchange a *shared secret* before they could start a secure communication session. Also, it was almost impossible to authenticate the user and stay anonymous. It's safe to say that the existence of bitcoin, and indeed many other computer systems, became possible for the first time back in 1977, when *asymmetric* or *public-key* encryption algorithms were first invented.

Unlike symmetric cryptography, asymmetric algorithms use different keys for encryption and decryption (Figure 5-4). One of the keys is kept in secret, and it's called a *private key*. Another key can be shared with anyone, and that's why it's called a *public key*.

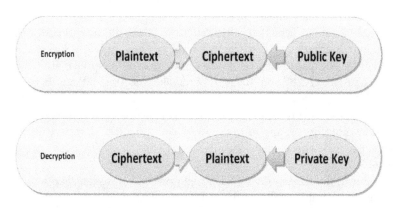

Figure 5-4: Public Key (Asymmetric) Encryption

There are several different implementations of public-key cryptosystems. Chapters 6 and 7 provide a comprehensive step-by-step explanation of two of them. RSA (Rivest, Shamir, Adleman) is used as an illustration of how public-key encryption works in general as well as being the first step to understanding the more complex elliptic curve cryptography (ECC), which is used in bitcoin implementation.

Bitcoin security rests on full trust in the power of public-key cryptography, which is the cornerstone of any cryptocurrency. There is a dilemma, which may affect the mainstream acceptance of bitcoin. On the one hand, if you don't understand the concept of public-key encryption, how can you trust cryptocurrency at all? But on the other hand, it is difficult to believe, at least at first glance, that someone without a degree in math can understand elliptic curves, which is the public-key encryption algorithm employed by bitcoin.

Digital Signatures

Cryptography is not limited to protection from information disclosure, otherwise we would not be talking about cryptocurrencies now because it would not be possible to implement them without another achievement of cryptography—digital signatures. Unlike encryption, which mostly takes care of confidentiality of the data, digital signatures ensure integrity and authenticity of transaction messages and records.

A credit card check digit is a very simplified "prototype" of a digital signature, which is intended to validate the integrity of the account number. Unfortunately, it does not fully preserve the integrity because everyone knows the mod 10 formula and can alternate the PAN and make it look like the original number by recalculating the check digit of new number. In addition, the check digit is unable to confirm the authenticity of the account number as it does not contain any information about the entity that performed the original check digit calculation. Let's try to resolve those problems.

From the description of hash functions earlier in this chapter, we know that simple replacement of mod 10 with a hash function such as SHA-256 does not resolve the problem because anyone can find out the SHA-256 formula and can easily tamper with the account number by recalculating the hash. The full solution can be achieved by adding public-key cryptography.

Let's encrypt this hash using a private key, and add this encrypted hash to the account number instead of mod 10. Now, anyone can validate our account number by calculating the hash, decrypting our encrypted hash using the public key (which we previously distributed to everyone), and comparing the two results. If the calculated hash is the same as the one that was originally encrypted, it means that the account number was never altered (Figure 5-5).

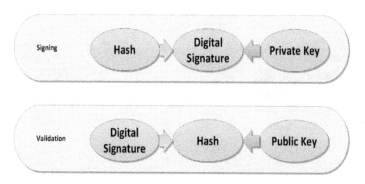

Figure 5-5: Digital Signature

Now no one can tamper with our account number because no one knows the private key! Also, anyone can confirm the authenticity of the account by using the public key, which is associated with a particular individual or organization—for example, the issuing bank. Since only this individual or organization has access to the private key, the fact that the decryption using the public key was successful means that the encryption was actually performed by the owner of the private key.

This is how digital signature algorithms work in general. Elliptic curve-based ECDSA, which is used for signing bitcoin transactions, is a little bit more complicated because in elliptic curve cryptography there is no straightforward way to encrypt data using the private key. However, the main principles remain the same, and if you understand public-key cryptography in general, and ECC in particular, you will be able to understand (and judge!) the level of bitcoin security. Therefore, the next three chapters are fully dedicated to understanding asymmetric encryption and elliptic curves.

CHAPTER 6

RSA Step by Step

All excellent things are as difficult as they are rare.
—Benedict Spinoza

One-Way Functions

If you ask about the main idea behind asymmetric encryption, the simple answer is that it is based on the difference in complexity of some complementary, opposite operations, or functions, when some calculation can be easily done in one direction, but it's very difficult to undo it, unless you have a secret back door —*private key*. Such operations are called *one-way*, or *trapdoor functions*. A good example of a one-way function is the relationship between programmers and mathematicians: most mathematicians know how to program; however, very few programmers are also good mathematicians.

Now let's move closer to our problem, and take, for example, two arithmetic operations—multiplication and *factoring*. Multiplication is a very basic operation that is even understood by programmers: $2 \times 3 = 6$ is a simple example of multiplication. Factoring is the opposite of multiplication, where complex numbers can be broken down to simple ones. In our example, 2 and 3 are *factors* of 6, or $6 = 2 \times 3$. Looking at this example, however, we don't see much difference in difficulty between multiplication ($2 \times 3 = 6$) and factoring ($6 = 2 \times 3$).

If you take very large numbers, multiplication still remains a relatively trivial operation that can be easily done by computer, calculator, or even with using pen and paper:

$619 \times 1373 = 849,887$ still does not look too terrible.

81

The factoring of a number of the same size, however, does not seem to be so easy:

```
905,741 = ? × ?
```

So how do we find the factors of this number? Programmers always first try the simplest way to solve their problem, unless there are limitations such as CPU or memory resources limits that force them to work on optimizing an existing algorithm or even to look for new one. Such an approach is also called brute-forcing. So the simplest, brute-force method of factoring is trying to match all possible combinations of factors:

6 = 2 × 2—wrong...

6 = 2 × 3—right!

In our first 6 = 2 × 3 example, the solution will be found just after the second integration of our algorithm, while finding the factors of 905,741 will take a significant number of iterations. Such an algorithm will have to work "forever" for really large numbers like this one:

```
13506641086599522334960321627880596993
88147560566702752448514385152651060485
95338339402871505719094417982072821644
71551373680419703964191743046496589274
25623934102086438320211037295872576235
85096431105640735015081875106765946292
05563685529475213500852879416377328533
90610975054433499981115005697723689092
756
```

Although there are other smarter mathematical methods of factorization, all of them require a lot of CPU time, which makes finding the result a computationally infeasible task for very large numbers (by the way, the answer for the previous primer is 905,741 = 587 × 1543 if you have not guessed it yet).

Let's Start

The RSA algorithm, which was one of the first practical implementations of public-key encryption, was invented by Ron Rivest, Adi Shamir, and Leonard Adleman in 1977.[54] RSA uses the multiplication of two *prime numbers* as a one-way function. A prime number, or *prime*, is any *natural number* that can be divided only by 1 or by itself without a remainder. (A natural number is a number that can be used for counting natural things like 1, 2, 3, etc.) Examples of a few "first" prime numbers are 2, 3, 5, 7, 11, 13, 17, 19, 23, 29, 31, etc. The fact that it's computationally infeasible to calculate the opposite function (factoring) of the result of multiplication of two large prime numbers defines the secrecy of the RSA system.

Now let's see how RSA encryption actually works. First of all, we need to generate a pair of public and private keys. RSA creates the public key first.

Public Key: Just a Random Number

In order to create a public key, the implementation generates a pair of random prime numbers and multiplies them. Let's take two small prime numbers—**11** and **13**—to make our example simple to understand and allow us to calculate and validate the results without a computer or even a calculator.

The result of multiplication of 11 and 13 is **143**, and it's called *modulus* (you will see why shortly). Now we can generate another random number, for example **7**, and call it a *public exponent*. In fact, the public exponent is our *public key*, which is used to encrypt the message, but usually it is not called a key because another part of the equation— the modulus—is required for the encryption.

Modulus: It's Like a Clock Dial

The modulus is needed to "limit" the results of multiplication during the encryption: when encrypting using a public exponent, the data (any number, for example 6) should be simply multiplied by itself public exponent (7) times. However, if the result of the multiplication is greater than the modulus, it is "truncated" to the closest number that is less than the modulus. Such an operation is called modulus, or *modulo*.

The best example that illustrates modulo is a clock dial (Figure 6-1). There are 24 hours in a day, but only 12 hours in a clock dial. Thus, the clock dial's modulus is 12. When we say 4 p.m., it actually means 16:00 (in military time format) because 4 is the result of modulo 12 operation on 16:

```
16 mod 12 = 4
```

Modulo is simply calculated by subtracting the modulus until the resulting number becomes less than the modulus (going one round counter clockwise):

```
16 mod 12 = 16 - 12 = 4
```

Figure 6-1: Clock Dial Shows 4:00 While Actual Time Is 16:00

Now, let's do a simple exercise: assuming the time now is 11:00 p.m. (23:00), we would like to set a timer to go off 20 hours from now. Whatever time format we use, we can take the current time plus the timer interval modulo 12 in order to get the target alarm setting:

```
11 + 20 mod 12 = 31 mod 12 = 31 - 12 - 12 =
7
```

```
23 + 20 mod 12 = 43 mod 12 = 43 - 12 - 12 - 12
= 7
```

However, how do we know whether the alarm should be set to 7 p.m. or 7 a.m. (19:00 or 7:00)? In order to know that, we actually calculate our time using modulo 24—the total number of hours in a day:

$$23 + 20 \bmod 24 = 19 \text{ (19:00 or 7 p.m.)}$$

We just learned some basics of *modular arithmetic* which is very handy when we get into the public-key cryptography calculations, as we will see later in this chapter.

Encryption: Plaintext to the Power of Public Key

Now let's encrypt our message (6) using public exponent (7) and modulus (143). We simply multiply the seven sixes: $6 \times 6 \times 6 \times 6 \times 6 \times 6 \times 6$ (6 to the power of 7, or 6^7). However, each time we step over 143 we go back to the closest number below 143:

Step 1: $6 \times 6 = 36$

Step 2: $36 \times 6 = 73$ (the result is actually 216, but remember, we cannot step over our modulus, which is 143. Imagine our clock dial has 143 hours, so we go back one round counterclockwise from 216: $216 - 143 = 73$.)

Step 3: $73 \times 6 = 9$ (it's actually 438 which is greater than 143, so we should "shrink" the result by going counterclockwise until it becomes less than 143: $438 - 143 = 295$; $295 - 143 = 152$; and finally: $152 - 143 = 9$)

Step 4: $9 \times 6 = 54$ (the result is already less than 143 so we are good here)

Step 5: $54 \times 6 = 38$ (go back counterclockwise twice: $324 - 143 \times 2 = 38$)

Step 6: $38 \times 6 = 85$ (truncate 228 just once: $228 - 143 = 85$)

The encrypted result is **85**.

Thus, the pair of the public exponent and modulus composes the RSA public key. In fact, if you look at the practical implementation of the format used to store the keys, such as PKCS #1, you can see that both public exponent and modulus numbers are present.[55]

In order to encrypt a longer message, for example, some text, it can be translated to series of numbers (ASCII characters), and then each number can be simply multiplied public exponent times, "adjusted" for modulus. In practice, however, the algorithm of the message encryption is more sophisticated because if you simply substitute the plaintext characters with encrypted characters, such ciphertext will be vulnerable to cryptanalysis, and it will be relatively easy to break such encryption.

Private Key: Phi Function + Modular Inversion

Don't be afraid of those two scary names in the section title; it's still pretty simple. In order to decrypt our encrypted message, we have to have the *private exponent*, which is calculated in advance using two methods (as in the case of the public exponent, the private exponent and the modulus together form the *private key*). The first formula is called *totient*, or *phi function*.[56] Its result is calculated using our initial pair of prime numbers, 11 and 13. In our example, the result of the phi function will be 120:

$$(11 - 1) \times (13 - 1) = 10 \times 12 = 120$$

The second step of finding the private exponent is called *modular inversion*, which is a bit more complicated, although for our small numbers it is still can be calculated even without a computer. However, since we are lazy, we can just use a special calculator (many of them are available online) and take the result.[57, 58] (If you are still interested in actual implementation of modular inversion, the C# and Python code are listed later in this chapter in a section called "Experimenting with the Code.") The input of the modular inversion calculator is two numbers: the public exponent and the result of the phi function, which we just found (7 and 120 in our example). The output is the value of the private exponent we are looking for: **103**.

Note that the pair of prime numbers, 11 and 13, which we initially multiplied to create the modulus 143, must remain secret as they allow one to calculate the private key by using them as part of the first step that finds the phi function (Table 6-1).

Table 6-1: RSA Parameters for Encryption and Decryption

RSA PARAM	NAME	OUR EXAMPLE	PUBLIC	PRIVATE
Pair of Prime Numbers	p and q	11 and 13	No	Yes
Modulus	n	143	Yes	Yes
Public Exponent	e	7	Yes	No
Private Exponent	d	103	No	Yes

Decryption: Ciphertext to the Power of Private Key

Now we can decrypt our message simply by multiplying the encrypted number (85) by itself the private exponent number of times (103), or 85 to the power of 103, or simply 85^{103}. But as with encryption, we should make sure to "shrink" the intermediate results as soon as they exceed the modulus value (143). The result of such decryption operation is the magic of RSA: we get our original number **6**:

$$85^{103} \bmod 143 = 6$$

Another side of this magic is that either public or private keys can be used for either encryption of decryption. In our previous example, we used the public key for encryption and the private key for decryption, which is common for asymmetric algorithms. The opposite scenario, when the private key is used to encrypt the message and the public key is used to decrypt it, is also possible with RSA. Normally, however, it does not make sense to encrypt with the private key so everyone can decrypt the message using the public key. The only situation where it is appropriate to encrypt with the private key is when using a *digital signature*. The digital signature scheme ECDSA, which is used in bitcoin implementation, is different from the one based on RSA. However, it is still based on the same principles used in all digital signatures.

CHAPTER 7

How Elliptic Curves Work

Our treasure lies in the beehive of our knowledge.
—Friedrich Nietzsche

Modern cryptography is mainly based on math. Obviously, the audience of this book is not primarily made up of mathematicians, so I'll try to explain the basics of elliptic curves cryptography (ECC) in layman's terms. I think the complexity of elliptic curves is an overestimated prejudice that we will try to overturn in this chapter. Anyone who has a clue about arithmetic and a basic knowledge of computers can understand how public-key encryption works.

Unlike RSA, it is difficult to explain elliptic curves without at least a single math formula. For this reason, we have reviewed the RSA in detail so you can get an idea of public-key encryption using something less complex. However, by doing it step by step, we will also break down the elliptic curves so that you can understand it without having knowledge of advanced mathematics.

The Graph

An elliptic curve is a function that can be represented graphically, as shown in Figures 7-1, 7-2, and 7-3, or in the form of an equation, $y^2 = x^3 + ax + b$, where x and y are horizontal and vertical coordinates of the points on the graph. By iterating through multiple values of horizontal coordinate x, we can easily calculate the matching values of vertical coordinate y, and thus we can draw the curve. By changing the values of a and b we can change the form of the curve. For example, a = –6 and b = 5 will produce the elliptic curve shown on Figure 7-1, while a = –1 and b = 1 will create a graph shown in

Figure 7-2. Bitcoin implementation[59] uses a curve with a = 0 and b = 7: $y^2 = x^3 + 7$, which is shown in Figure 7-3.

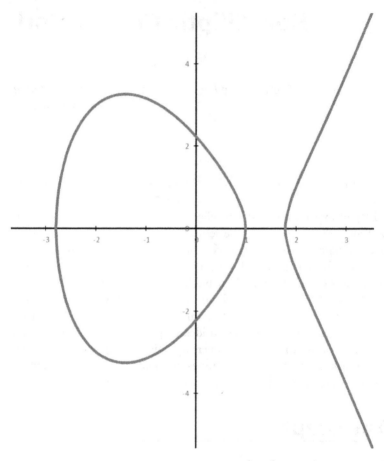

Figure 7-1: Example of an Elliptic Curve: $y^2 = x^3 - 6x + 5$

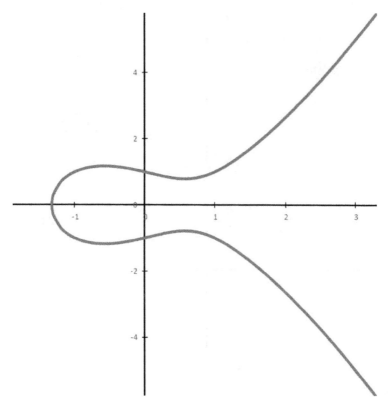

Figure 7-2: Another Example of an Elliptic Curve: $y^2 = x^3 - x + 1$

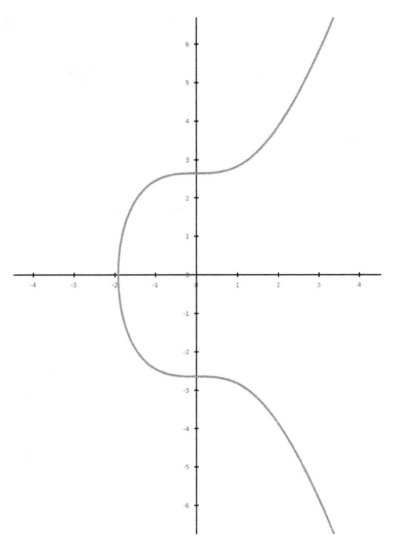

Figure 7-3: Bitcoin Elliptic Curve: $y^2 = x^3 + 7$

Horizontal Symmetry and Points of Intersection

One of the interesting features of any elliptic curve is its symmetry with respect to the horizontal axis: The part of the graph below the x-axis is mirroring the part above the x-axis (Figure 7-4). So any point with coordinates (x, y) has a corresponding point (x, −y). Let's see how this *horizontal symmetry* is used in so-called *point operations*. In order to do that, let's first review other important features of elliptic curves.

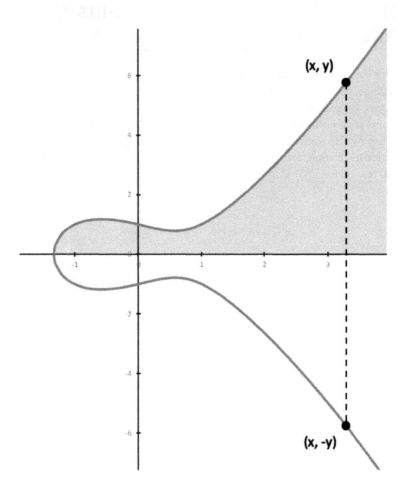

Figure 7-4: Horizontal Symmetry

It turns out that if you draw a straight line that intersects the elliptic curve at two points (let's say points A and B), in most cases there will be another point of intersection (Figure 7-5).

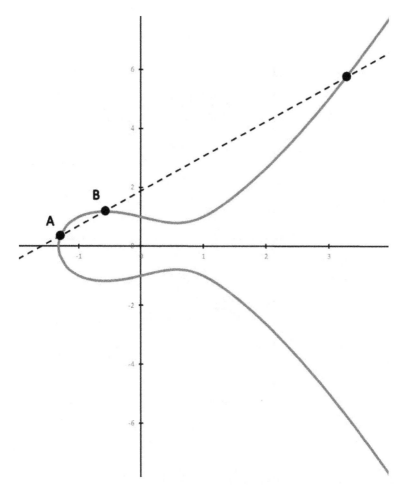

Figure 7-5: 3-Point Intersection with a Straight Line

However, there will be an "exception" when you draw a tangent line, which just touches the curve at single point A. In this case, there will be only two points where the tangent line intersects with the curve: the point A itself and another point (Figure 7-6).

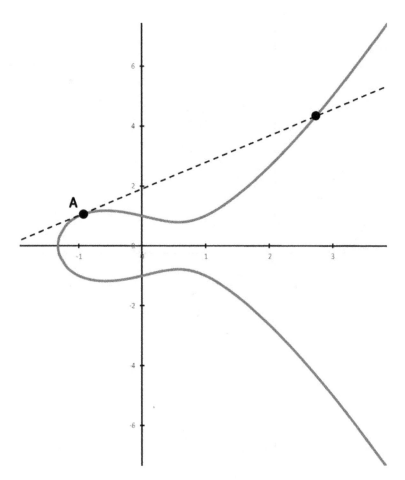

Figure 7-6: 2-Point Intersection with a Tangent Line

Point Operations

These three features of elliptic curves—horizontal symmetry, three-point intersection with a straight line, and two-point intersection with a tangent line—are used to form the two very important basic point operations: *point addition* and *point doubling*. Since doubling is in a way a subset of addition, let's review the point addition first and then the point doubling, so the latter will be clear as soon as we understand the former.

Point Addition

Point addition is based on two out of three features: three-point intersection with a straight line and horizontal symmetry. Let's just slightly extend the picture shown in Figure 7-6 and draw the vertical line through the third point of intersection. This line will intersect the mirroring point C below the x-axis (Figure 7-7). This point C is the result of point addition of A and B:

```
C = A + B
```

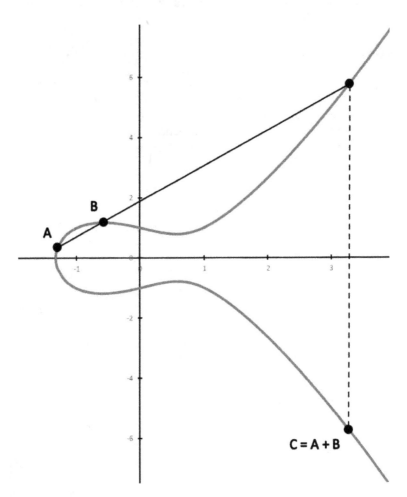

Figure 7-7: Point Addition

Point Doubling

Point doubling can be done in exactly the same way by drawing a vertical line from the second point of intersection. It will also reach the "mirroring" part of the elliptic curve at point C, which is the result of point doubling of point A (as shown in Figure 7-8):

```
C = 2A
```

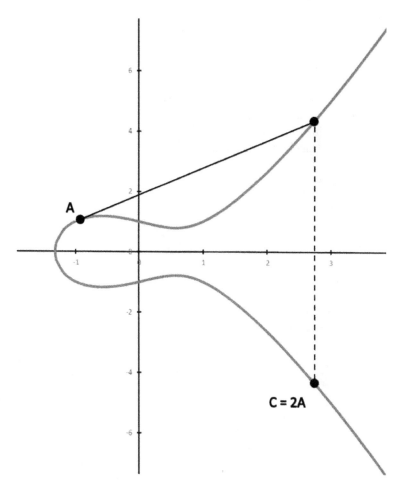

Figure 7-8: Point Doubling

You may be already confused and asking "why do we need all these operations, and how is this related to cryptography?" Don't worry, the answer is already close.

Point Multiplication

Now, when we know to add up points (including adding the point to itself by doubling), we can take virtually any point and add it to itself multiple times. The result will always be another point on the curve.

An example in Figure 7-9 shows point D, which is point A added to itself twice by first doubling A (Figure 7-8) and then adding the result of doubling—point C—to initial point A:

```
C = 2A
```

```
D = C + A = 2A + A = 3A
```

So, we just introduced another operation: *point multiplication*, which is also called *scalar point multiplication*.

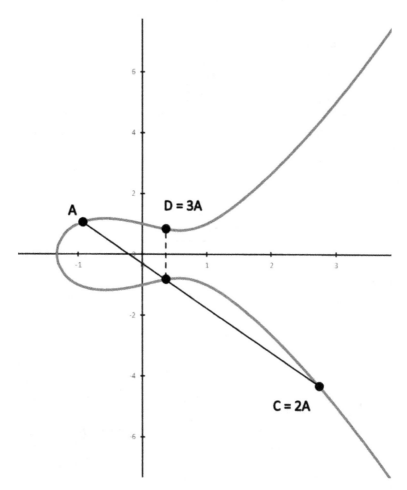

Figure 7-9: Point Multiplication

Now imagine that you multiply a point a thousand times, or a million times, or n times, where n is a very large number. By using a combination of just two basic point operations—addition and doubling—it is relatively easily for computer to multiply point A by a very large number n and get the resulting point L = nA. Using special *double-and-add algorithms*, we can efficiently combine point addition and doubling operations in order to find a shortest way to the result of scalar multiplication. For example, in order to calculate L = nA with *n*=9, the intuitive approach would be adding A to itself eight times:

```
L = 9A = A + A + A + A + A + A + A + A +
A
```

However, such a method would not be efficient with very large number *n*. Using the double-and-add algorithm, the same calculation can be performed using only four operations instead of eight: three point doublings and one point addition:

```
L = 9A = 2(2(2A))) + A
```

One-Way Function

Now here is the most important part of this chapter. Even if you know the coordinates of the initial point A and the final point *L*, it is very difficult, almost impossible, or *computationally infeasible* to reproduce the exact sequence of all those multiple point additions and doublings in order to determine *n*—the number of additions of the point A to itself that led to the resulting point *L*. Doesn't it already sound like a one-way function? Yes, it does. This is in fact a one-way function that is called the *elliptic curve discrete logarithm problem*—the foundation of elliptic curve cryptography,[60] which was discovered in 1985 by Victor Miller and Neil Koblitz.[61]

Now let's see how the elliptic curve discrete logarithm problem can be transformed into an actual working cryptosystem. In order to get closer to a little or less acceptable practical application, we need to do at least three things.

First, we need to transfer our one-way function from the "visual" graphical world to the "hidden" world of numbers, which is more convenient for computing, which means providing ourselves with elementary *algebraic* or even simpler *arithmetic* formulas so we can efficiently calculate all those point additions, doublings, and multiplications without the need to draw the lines. Fortunately for us, mathematicians, as usual, have already done all the hard work. There are simple if not relatively straightforward formulas available for us to use to calculate C = A + B and C = 2A. (As we agreed not to litter the text with math equations, we will omit them for now, but we will still have to see them later in order to observe the encryption in action.)

Second, it would be difficult to use elliptic curves with real numbers in practical cryptographic implementation because the calculations with real numbers are slow, and the results of operations with real numbers are inaccurate due to rounding associated with the real number arithmetic. Since we deal with computers, which prefer to crunch the simple and precise natural numbers rather than complex and inaccurate real ones, we need to limit ourselves to the points with x- and y-coordinates that are natural numbers. In order to do that, we iterate through the x-axis, taking only natural numbers for horizontal coordinate x and calculating the resulting coordinates y, which are also going to be a natural numbers. The result of such drawing will not be as nice looking as previous graphs because it will consist of just a bunch of points with "natural" coordinates only (like 1, 2, 3, 4, etc.), which are not connected with each other (Figure 7-10). But such a picture is acceptable for us at this point because this is the last time we actually see the elliptic curve as a graph in this book—we don't need them anymore as all we need to continue is the algebraic formulas "connecting the dots."

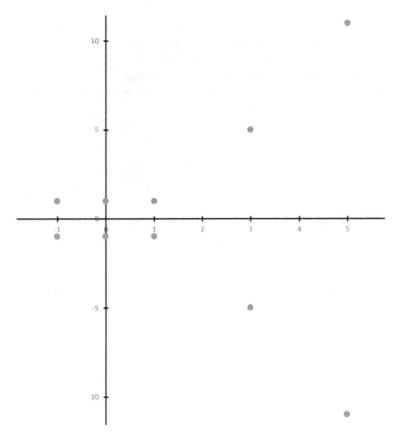

Figure 7-10: Elliptic Curve $y^2 = x^3 - x + 1$ with Points Having Only Natural Numbers as Their Coordinates

And finally, by learning from other asymmetric cryptosystems, we know that despite the fact that we have to operate with large numbers in order to make the system secure, it is impossible to deal with very large numbers produced by the elliptic curve equation, so we have to limit our "working zone" by modulus.

Limiting the Curve for the Sake of Cryptography

Remember the RSA modulus? The concept is the same: we limit both x and y values to some prime number p, so the new formula of the elliptic curve will look like $y^2 \bmod p = (x^3 + ax + b) \bmod p$. Figure 7-11 shows the same elliptic curve $y^2 = x^3 - x + 1$ rendered over a *finite field* with modulus 17. You can see that the curve, or rather what is left of it, even keeps some visual horizontal symmetry around the axis located between x = 8 and 9.

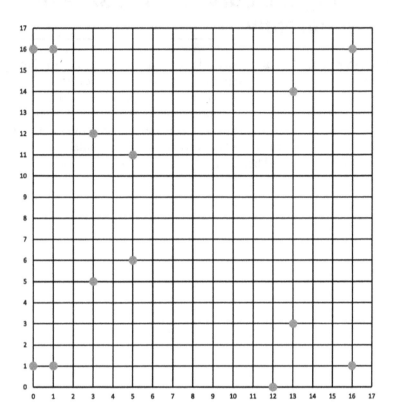

Figure 7-11: Elliptic Curve over Finite Field with Modulus 17

It's important to note that all the point operations—addition, doubling, and scalar multiplication—still can be calculated on such a heavily modified elliptic curve. However, the values will always be inside the field, that is, between zero and the value of the modulus. When we are finally able to operate with whole numbers only and within the predefined range (and that's all computers need), we are ready to implement the actual crypto system.

Generating the Keys

Unlike RSA, where the public key is generated first, the public key in the ECC system is derived from the private key, which is just a random number. You probably already guessed what its value means—the number of times the initial point on the curve is "scrambled" with itself using point multiplication, which we already discussed earlier.

In order to calculate the public key (let's call it Q), we need to select an initial point on our curve (let's call it P) and multiply it by private key d:

$$Q = dP$$

Note that points in our formulas are always represented by capital letters like A, B, C, D, Q, L, and P, while whole numbers are represented by lowercase letters such as d, k, and n. Points can also be defined as a pair of whole numbers—x- and y-coordinates: (x, y).

Encryption

Now we are ready to encrypt the plaintext, which must be represented by any point M on our curve. The real-world implementations of encryption with elliptic curves, which consist of very large numbers of points, use special algorithms to convert any plaintext into the point on the curve. Also, such implementations often use a *hybrid* approach when the plaintext is encrypted using symmetric algorithms. In our simplified example we will use an encryption based on the ElGamal scheme.[62]

In order to encrypt the message, we take the public key Q of the recipient, select a random number k, and calculate the ciphertext, which consists of two points, C1 and C2:

$$C1 = kP$$

$$C2 = M + kQ$$

So the entire data that's being sent by the sender who encrypts the message consists of just C1 and C2. The public key is known to everyone, and the private key is known only to the recipient.

Decryption

Upon receipt of the ciphertext (points C1 and C2), the recipient will use the private key d to decode it into the original message M using the following formula:

```
M = C2 - dC1
```

Isn't it simple and elegant? Well, maybe it's elegant but not so simple when you proceed to doing calculations with actual formulas and numbers.

Just a Little Bit of Math

Now it's time to complicate our lives a little bit and bring to life the formulas of point addition, doubling, and multiplication, along with their implementation, and operate with real numbers rather than just play with letters. As we agreed to not overload the text with math, you don't have to pay attention to the formulas if you don't want to; just believe that eventually they work. You can discard the whole process of calculation and go directly to the next chapter. In order to make those equations more attractive for developers, however, there are also code samples in C# available for each formula.

Note that these code samples are not intended to be real code but were created only to illustrate the process of calculation for those who are more familiar with seeing source code rather than mathematical formulas.

Point Addition: C = A + B

Following are the formulas used for calculation of the coordinates x_c and y_c of point C, which is the result of point addition of points A and B with coordinates (x_a, y_a) and (x_b, y_b):

$$s = \frac{y_b - y_a}{x_b - x_a} \bmod p$$

$$x_c = (s^2 - x_a - x_b) \bmod p$$

$$y_c = (s(x_a - x_b) - y_a) \bmod p$$

Note that coefficient s is calculated separately because it is used for computing both x- and y- coordinates.

Point Doubling: C = A + A = 2A

$$s = \frac{(3x_a)^2 + a}{2y_a} \bmod p$$

$$x_c = (s^2 - 2x_a) \bmod p$$

$$y_c = (s(x_a - x_c) - y_a) \bmod p$$

Note that the formula of coefficient s is different, and includes a, which is the value of coefficient a from the elliptic curve equation $y^2 = x^3 + ax + b$. (Do not confuse this with point A and its coordinates x_a and y_a.) In our sample calculations that follow, we use a = −1 as our curve is $y^2 = x^3 - x + 1$.

Now Let's Play with the Numbers

In our simplified example, let's select an initial point on our curve with coordinates x = 3 and y = 5, and choose 5 as the private key value:

d = 5

P = (3, 5)

The coordinates of the public key (which is also a point on the curve!) can then be calculated using point multiplication:

Q = 5P

The process can be visualized as shown in Figure 7-12. However, in practice, multiplication is implemented via a combination of doublings and additions.

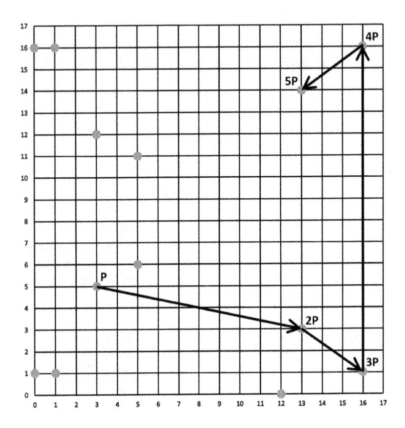

Figure 7-12: Finding the Public Key

Since the value of d is very small, we can "manually apply" the double-and-add algorithm, replacing the multiplication by 5 with two point doublings and one addition (Figure 7-13):

```
Q = 5P = 4P + P = 2(2P) + P
```

If you are interested in actual implementation of the double-and-add algorithm, it is described in chapter 8.

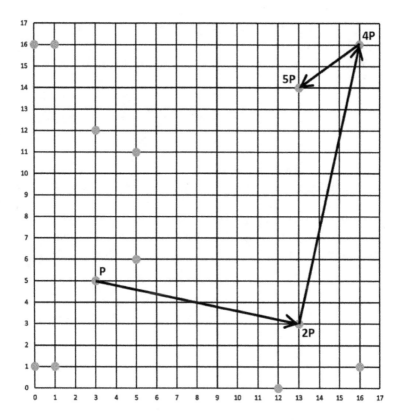

Figure 7-13: Calculating the Public Key Using the Double-and-Add Algorithm

Now, using the formulas shown previously, we can calculate the public key Q based on the coordinates of initial point P = (3, 5) and the value of private key d = 5:

```
Q = 5 × (3, 5) = 2 × (2 × (3, 5)) + (3,
5)
```

First, let's calculate 2P = 2 × (3, 5) using point doubling formulas:

$s = ((3 \times 3^2 + -1) / 2 \times 5) \bmod 17 = (26 \bmod 17 \times 1 / 10 \bmod 17) \bmod 17 = (9 \times 12) \bmod 17 = 108 \bmod 17 = 6.$

$x_c = (6^2 - 2 \times 3) \bmod 17 = 30 \bmod 17 = 13.$

$y_c = (6 \times (3 - 13) - 5) \bmod 17 = -65 \bmod 17 = 3.$

So the result of 2P is (13, 3).

Now, using the same calculation method we can find that $2 \times 2P = 2 \times (13, 3) = (16, 16)$.

Finally, let's perform the last operation—point addition—in order to find out the value of the public key Q: using the preceding point addition formulas, add the result of two doublings (16, 16) to the initial point (3, 5):

$s = ((5 - 16) / (3 - 16)) \bmod 17 = (-11 \bmod 17 \times -1 / 13 \bmod 17) \bmod 17 = (6 \times 13) \bmod 17 = 10$

$x_c = (10^2 - 16 - 3) \bmod 17 = 81 \bmod 17 = \mathbf{13}$

$y_c = (10 \times (16 - 13) - 16) \bmod 17 = 14 \bmod 17 = \mathbf{14}$

So here is the value of our public key:

$\mathbf{Q} = 5 \times (3, 5) = 2 \times (13, 3) + (3, 5) = (16, 16) + (3, 5) = \mathbf{(13, 14)}.$

The public key Q, along with other parameters of the curve, can be exposed to anyone who wants to encrypt the message.

Encryption

In our simplified example there are a limited number of points on the curve—only 13—so it's difficult to represent any arbitrary plaintext as a point on the curve. Let's improvise and imagine that we would like to only encrypt number **5**. Then we need to find point M on our curve with horizontal coordinate x = 5. Fortunately, there is such a point: (5, 6). Based on the formulas previously listed, the resulting ciphertext will consists of two points, C1 and C2, where C1 = kP and C2 = M + kQ, and k is just a random number. Let's pick k = 3 and calculate C1 first using point multiplication (which is, in fact, replaced with one doubling and one addition):

```
C1 = 3 × (3, 5) = 2 × (3, 5) + (3, 5)
   = (13, 3) + (3, 5) = (16, 1).
```

C2 is the result of addition of point M to the result of the public key Q multiplied by k:

```
C2 = (5, 6) + 3 × (13, 14) =
(5, 6) + 2 × (13, 4) + (13, 14) =
(5, 6) + (16, 1) + (13, 14) =
(5, 6) + (3, 5) = (5, 11)
```

Decryption

In order to decrypt the ciphertext C1, C2, one needs to know the private key d. All other parameters are public (Table 7-1). According to the formulas already described, M= C2 − dC1:

```
M = (5, 11) − 5 × (16, 1)
```

We don't know how to subtract points, but we can multiply by a negative number, so let's multiply point (16, 1) by −5 and then add the result to point (5,11):

```
M = (5, 11) + (−5 × (16, 1)) =
(5, 11) + (3, 12) = (5, 6)
```

The result is our initial message: point with coordinates x = 5 and y = 6, and we get back to our initial plaintext **5**. Elliptic curves cryptography works!

Table 7-1: ECC Parameters for Encryption and Decryption

ECC PARAMETER	CONVENTIONAL NAME (MAY VARY)	OUR EXAMPLE	EXPOSED TO ENCRYPTION SIDE (PUBLIC)	IN USE BY DECRYPTION SIDE (PRIVATE)
Coefficients that define the curve	*a and b*	−1 and 1	Yes	Yes
Initial point	*P*	(3, 5)	Yes	Yes
Modulus	*p*	17	Yes	Yes
Public Key	*Q*	(13, 14)	Yes	No
Private Key	d	5	No	Yes
Random Number	k	3	Yes	No

Experimenting with
the Code

This short chapter is the last resort intended for "nonbelievers"—those who want to make sure that the results in the example are not taken out of thin air and see how the numbers were actually computed. Just skip it if you are not programmer.

Those formulas in chapter 7 look relatively simple but there are several "tricks" required in order to implement their computation in practice. You can see that all the operations end with *mod p*. That's because all those operations relate to *modular arithmetic*. Some languages, including C#, do not have built-in modular arithmetic operators, so some extra work is required in order to implement our point operations. Also, note that the functions that calculate the large-size elliptic curves, which are suitable for real-world cryptography, need to operate with *BigInteger* rather than just regular-size *integer* types, so *int* type in both C# and Python code samples is used for the sake of simplicity and clarity. The full source code of those samples can be download from the book page on my website http://www.gomzin.com.

And again, if you are not a coder or not interested in seeing the source code at all, skip this section and go to the next "codeless" one about digital signatures.

Modulus

There is a C# operator for calculating the modulus: %. In order to calculate the value of *a mod p*, you just write this line: b = a % p. However, this code may return a wrong result if a has a negative value because our curve is limited to only positive coordinates of the points. We need to correct the result of the standard modulus, and so the full code in C# would be something like this:

```csharp
int Mod(int a, int p)
{
    int b = a % p;
    if (b < 0)
        b = (b + p) % p;
    return b;
}
```

The Python code looks similar:

```python
def Mod(a, p):
    b = a % p
    if b < 0:
        b = (b + p) % p
    return b
```

Modular Inversion

Modular inversion is required when the modulus operation is performed on $\frac{1}{a}$ instead of a:

$$b = \frac{1}{a} \bmod p = a^{-1} \bmod p.$$

We will use it for calculating the s coefficients for point addition and doubling. Note that the same algorithm/code was used as part of the RSA private key calculation. Here is the C# code for modular inverse computation:

```csharp
int ModInverse(int a, int p)
{
    int i = p;
    int v = 0;
    int d = 1;
    a = Mod(a, p);
```

```
    while (a > 0)
    {
        int t = i / a;
        int x = a;
        a = i % x;
        i = x;
        x = d;
        d = v - t * x;
        v = x;
    }
    v %= p;
    if (v < 0)
        v = (v + p) % p;
    return v;
}
```

The same algorithm in Python:

```
def ModInverse(a, p):
    i = p
    v = 0
    d = 1
    a = Mod(a, p)
    while a > 0:
        t = int(i / a)
        x = a
        a = i % x
        i = x
        x = d
        d = v - t * x
        v = x
        v %= p
    if v < 0:
        v = (v + p) % p
    return v
```

Representing the Points

Before we go to the point operations, let's introduce a simple point class that encapsulates the coordinates x and y and basic comparison operations with points:

```
class Point
{
    public int x;
    public int y;

    public Point() : this(0, 0) { }

    public Point (int X, int Y)
    {
        x = X;
        y = Y;
    }

    public bool IsInfinity
    {
        get
        {
            return x == 0 && y == 0;
        }
    }

    public bool IsEqual(Point A)
    {
        Return this.x == A.x &&    this.y ==
A.y;
    }
}
```

Point Doubling

Now that we know how to compute the modulus and modular inversion, nothing can prevent us from calculating point doubling and addition. For either point addition or doubling, first we find out the value of coefficient s.

Here are the functions that return the value of s for point doubling in C#:

```
int s_doubling(Point A, int p, int a)
{
    Int upper = Mod (3 * (int)
Math.Pow(A.x, 2) + a, p);
    int lower = modInverse(2 * A.y,
p);
    int s = Mod(upper * lower, p);
    return s;
}
```

And Python:

```
def s_doubling(A, p, a):
    upper = Mod(3 * A.x ** 2 + a, p)
    lower = ModInverse(2 * A.y, p)
    s = Mod(upper * lower, p)
    return s
```

Now, using coefficient s, we can perform the actual doubling:

```
Point Double(Point A, int p, int a)
{
    Point C = new Point();
    int s = s_doubling(A, p, a);
    C.x = Mod((int)Math.Pow(s, 2) - 2
*  A.x, p);
    C.y = Mod(s * (A.x - C.x) - A.y,
p);
    return C;
```

```
}
```

Python:

```python
def Double(A, p, a):
    C = Point()
    s = s_doubling(A, p, a)
    C.x = Mod(s ** 2 - 2 * A.x, p)
    C.y = Mod(s * (A.x - C.x) - A.y,
p)
    return C
```

Point Addition

Point addition follows the same pattern: first, calculating coefficient s, and then the addition algorithm itself:

```c
int s_addition(Point A, Point B, int
p)
{
    int upper = Mod(B.y - A.y, p);
    int lower = ModInverse(B.x - A.x,
p);
    int s = Mod(upper * lower, p);
    return s;
}

Point Add(Point A, Point B, int p, int
a)
{
    if (A.IsInfinity)
        return B;
    if (B.IsInfinity)
        return A;
    if (A.IsEqual(B))
        return Double(A, p, a);
    Point C = new Point();
```

```
    int s = s_addition(A, B, p);
    C.x = Mod((int)Math.Pow(s, 2) -
A.x - B.x, p);
    C.y = Mod(s * (A.x - C.x) - A.y,
p);
    return C;
}
```

Point addition in Python:

```python
def s_addition(A, B, p):
    upper = Mod(B.y - A.y, p)
    lower = ModInverse(B.x - A.x, p)
    s = Mod(upper * lower, p)
    return s



def Add(A, B, p, a):
    if A.IsInfinity():
        return B
    if B.IsInfinity():
        return A
    if A.IsEqual(B):
        return Double(A, p, a)
    C = Point()
    s = s_addition(A, B, p)
    C.x = Mod(s ** 2 - A.x - B.x, p)
    C.y = Mod(s * (A.x - C.x) - A.y,
p)
    return C
```

Point Multiplication

Finally, point multiplication can be done by combining additions and doublings—the *double-and-add algorithm*:

```
Point Multiply(Point A, int d, int p, int a)
{
    Point R = new Point(0, 0);
    Point Q = A;
    while (d > 0)
    {
        int t = d % 2;
        if (t == 1)
        {
            R = Add(R, Q, p, a);
        }
        Q = Double(Q, p, a);
        d = d / 2;
    }
    return R;
}
```

In Python:

```python
def Multiply(A, d, p, a):
    R = Point()
    Q = A
    while d > 0:
        t = d % 2
        if t == 1:
            R = Add(R, Q, p, a)
        Q = Double(Q, p, a)
        d = int(d / 2)
    return R
```

Calculating the Public Key

After we have all the code, the calculation of the public key is as simple as the following lines of code.

Define the curve:

```
int p = 17,   a = -1;
```

```
Choose the Private Key:
int d = 5;
Calculate the Public Key:
Point P = new Point(3, 5);
Point Q = Point.Multiply(P, d, p, a);
```

Python:

```
p, a = 17, -1
d = 5
P = Point(3, 5)
Q = Multiply(P, d, p, a)
```

The result is **(13, 14)**.

Encryption

Encryption will look simple as well.

Select the plaintext represented by any point M on the curve, and random number k:

```
Point M = new Point(5, 6);
int k = 3;
```

Calculate the ciphertext, which consists of two points, C1 and C2:

```
Point C1 = Multiply(P, k, p, a);
Point kQ = Multiply(Q, k, p, a);
Point C2 = Add(M, kQ, p, a);
```

Following is the resulting ciphertext in our example:

```
(16, 1) (5, 11)
```

We get the same result in Python:

```
M = Point(5, 6)
```

```
k = 3
C1 = Multiply(P, k, p, a)
kQ = Multiply(Q, k, p, a)
C2 = Add(M, kQ, p, a)
```

Decryption

In order to decrypt the ciphertext C1, C2, we use the private key d:

```
Point dC1 = Multiply(C1, d, p, a);

Point Minus_dC1 =
new Point(dC1.x, Mod(-dC1.y, p));

Point M_Decrypted =
Add(C2, Minus_dC1, p, a);
```

The result of the preceding calculation is **(5, 6)**, the coordinates of our original plaintext point M.

The result is the same in Python:

```
dC1 = Multiply(C1, d, p, a)
Minus_dC1 = Point(dC1.x, Mod(-dC1.y,
p))
M_Decrypted = Add(C2, Minus_dC1, p, a)
```

Part II Summary

Cryptography is an essential part of bitcoin implementation. While *symmetric encryption* algorithms are not used in bitcoin, *one-way hash functions* and *public-key cryptography* together compose the *digital signature* mechanism, which protects the *integrity* and provides the *authenticity* of the crypto transactions.

SHA-256 and *RIPEMD-160* are the two one-way hash function algorithms used in bitcoin implementation. *ECDSA*, which is based on a public-key cryptosystem called *elliptic curve cryptography*, is used to digitally sign the bitcoin transactions.

Public-key cryptography is essential for the security of any cryptocurrency. Bitcoin implementation employs *elliptic curve cryptography* (ECC).

An elliptic curve is a function in a form $y^2 = x^3 + ax + b$, where x and y are horizontal and vertical coordinates. ECC is based on the unique features of elliptic curves:

- horizontal symmetry
- three-point intersection with a straight line
- two-point intersection with a tangent line

Using those features, three operations with the points on the elliptic curve can be defined:

- point addition: $C = A + B$
- point doubling: $C = 2A$
- scalar point multiplication: $C = kA$

Point multiplication is implemented as a combination of point additions and doublings, which is determined by the *double-and-add algorithm*. For example, $3A = 2A + A$.

Using scalar point multiplication we can "multiply" an *initial point* on curve P by very large random number d. This number d is a *private key*, while the result of multiplication Q is a *public key* in the ECC system: $Q = dP$. It is *computationally infeasible* to reverse this operation without the knowledge of the private key, which forms a *one–way function* called the *elliptic curve discrete logarithm problem*.

Practical implementations of ECC use elliptic curve over a *finite field*—the curve with the points represented by natural numbers only, and limited to some very large number, *modulus*. All operations with such a "limited" elliptic curve involve modular arithmetic.

The encryption/decryption example uses a simplified ECC algorithm based on the ElGamal scheme. The plaintext is represented by point M on the elliptic curve. The resulting ciphertext consists of two points, C1 and C2.

The examples of source code in C# and Python illustrate some basic ECC calculations. The real-world implementation of ECC operates with big integer types.

References

1. Ab ovo (Latin): from the beginning, from the origin, from the egg

2. Adam Smith, *An Inquiry into the Nature and Causes of the Wealth of Nations*, Reprint edition (New York: Bantam Classics, 2003).

3. Murray N. Rothbard, *The Mystery of Banking*, (New York: Richardson & Snyder, 1983), 7.

4. Slava Gomzin, *Hacking Point of Sale: Payment Application Secrets, Threats, and Solution* (Hoboken, NJ: Wiley, 2014).

5. "Smart card wallet," Bitcoin wiki, January 2013, https://en.bitcoin.it/wiki/Smart_card_wallet.

6. Loop, 2015, http://www.looppay.com.

7. Mobile Apps, Starbucks, 2015, http://www.starbucks.com/coffeehouse/mobile-apps.

8. Slava Gomzin, "Mobile Checkout: Secure Mobile Payments," April 2009, http://www.gomzin.com/mobile-checkout.html.

9. Sergey P. Kapitza, *Global Population Blowup and After: The Demographic Revolution and Information Society*, (Russian Academy of Sciences, 2006), http://citeseerx.ist.psu.edu/viewdoc/download?doi=10.1.1.132.3287&rep=rep1&type=pdf.

10. Sergey Kapitza was a Russian physicists and demographer, born in Cambridge, England in 1928. He is known for his work in the areas of applied electrodynamics and accelerator physics, as well as historical demography. Kapitza created the theory of hyperbolic population growth and global demographic transition.

11. Jack Weatherford, *The History of Money* (New York: Three Rivers Press, 1997).

12. Aleksey Nikolayevich Tolstoy (1883–1945) was a Russian writer. Do not confuse him with Leo Tolstoy; there are three very well-

known writers in Russia with the same last name: Aleksey Konstantinovich Tolstoy, Lev "Leo" Nikolayevich Tolstoy, and Alexei Nikolayevich Tolstoy. Only Leo Tolstoy gained international fame for his great novels *War and Peace* and *Anna Karenina*.

13. Alexei Tolstoy, *The Garin Death Ray*, First revised edition (Foreign Languages Press, 1955).

14. Nick Taylor, *Laser: The Inventor, the Nobel Laureate, and the Thirty-Year Patent War* (New York: Simon & Schuster, 2000).

15. Willem Middelkoop, *The Big Reset: Gold Wars and the Financial Endgame* (Amsterdam : Amsterdam University Press, 2014).

16. "Feds Accuse E-Gold of Helping Cybercrooks," *NBC News*, May 2, 2007, http://redtape.nbcnews.com/_news/2007/05/02/6346006-feds-accuse-e-gold-of-helping-cybercrooks.

17. e-gold Shopping Cart Interface, 25 January, 2001, e-gold Ltd., http://web.archive.org/web/20010126095200/http://www.e-gold.com/docs/e-gold_sci.html.

18. CRYPTOAdmin SPT Authentication Server v5.32 Administrator Guide, CRYPTOCard Corporation, 2003, http://portal.cryptocard.com/documentation/TechDocs/CRYPTOAdminV532.pdf.

19. Felix Martin, *Money: The Unauthorized Biography*, (New York: Alfred A. Knopf, 2014), 27.

20. OnlinePaySystems.INFO, http://www.onlinepaysystems.info.

21. Dr. David Chaum is the founder of DigiCash, a company that has pioneered electronic cash innovations. In the area of cryptography, he has published over 45 original technical articles. Website: http://www.chaum.com.

22. David Chaum, "Blind Signatures for Untraceable Payments," *Advances in Cryptology: Proceedings of Crypto 82*, (New York: Springer, 1983), http://link.springer.com/chapter/10.1007%2F978-1-4757-0602-4_18.

23. David Chaum, Amos Fiat, Moni Naor, "Untraceable Electronic Cash," *Advances in Cryptology — CRYPTO ' 88*, (New York: 1990), http://link.springer.com/chapter/10.1007%2F0-387-34799-2_25.

24. "The Overview of E-cash: Implementation and Security Issues," Global Information Assurance Certification Paper, SANS Institute, 2002, http://www.giac.org/paper/gsec/1799/overview-e-cash-implementation-security-issues/103204.

25. "How DigiCash Blew Everything," *NEXT*, January 1999, http://cryptome.org/jya/digicrash.htm.

26. Lev Grossman, "Beenz Counters," *Time*, February 2000, http://content.time.com/time/subscriber/article/0,33009,9961 11,00.html.

27. Mark W. Vigoroso, "Beenz.com Closes Internet Currency Business," *E-Commerce Times*, August 17, 2001, http://www.ecommercetimes.com/story/12892.html.

28. Jane Martinson, "Flooz.com Expires after Suffering $300,000 Sting," *The Guardian*, August 28, 2001, http://www.theguardian.com/technology/2001/aug/28/newme dia.business.

29. "Notice of Finding that Liberty Reserve S.A Is a Financial Institution of Primary Money Laundering Concern," Department of the Treasury, May 2013, p. 4, http://www.fincen.gov/statutes_regs/files/311--LR-NoticeofFinding-Final.pdf.

30. "Full Text of 'Liberty Reserve Indictment,'" *The Internet Archive*, https://archive.org/stream/704540-liberty-reserve-indictment/704540-liberty-reserve-indictment_djvu.txt.

31. Liberty Reserve domain seized by law enforcement agencies, http://libertyreserve.com.

32. BitTorrent is a technology that allows one to share and transfer files of just about any size quickly and efficiently. It works by breaking files up into small pieces. The file is downloaded piece by piece from one or many different sources. It's efficient

because you get faster downloads using a lot less bandwidth. More information can be found at http://www.bittorrent.com/help/faq/concepts.

33. The list of seized domains associated with Liberty Reserve, The US Department of Justice, http://www.justice.gov/usao/nys/pressreleases/May13/Liberty ReserveetalDocuments/Liberty%20Reserve,%20et%20al.%20Re lated%20Exchanger%20Website%20Domain%20Names%20Re dacted%20Filed%20Complaint%2013CV3565%20final%20with %20exhibits.pdf.

34. Use PayPal to pay in store, *PayPal*, https://www.paypal.com/webapps/mpp/pay-in-stores.

35. Dan Geer, "Cybersecurity as Realpolitik," Keynote for Black Hat USA 2014 conference in Las Vegas, August 6, 2014, http://geer.tinho.net/geer.blackhat.6viii14.txt.

36. Satoshi Nakamoto, "Bitcoin: A Peer-to-Peer Electronic Cash System," Bitcoin.org, (2008), https://bitcoin.org/bitcoin.pdf.

37. LikeInAMirror (blog), "Occam's Razor: Who Is Most Likely to Be Satoshi Nakamoto?," March 2014, https://likeinamirror.wordpress.com/2014/03/11/occams-razor-who-is-most-likely-to-be-satoshi-nakamoto/comment-page-1/.

38. Leah McGrath Goodman, "The Face Behind Bitcoin," *Newsweek*, March 2014,.

39. Andy Greenberg, "New Clues Suggest Craig Wright, Suspected Bitcoin Creator, May Be a Hoaxer," *Wired*, December 2015, http://www.wired.com/2015/12/new-clues-suggest-satoshi-suspect-craig-wright-may-be-a-hoaxer.

40. Andy Cush, "The Strange Life and Death of Dave Kleiman, A Computer Genius Linked to Bitcoin's Origins," *Gizmodo*, December 2015, http://gizmodo.com/the-strange-life-and-death-of-dave-kleiman-a-computer-1747092460.

41. Satoshi Nakamoto, "Bitcoin P2P e-cash paper," *Gmane*, October 2008,

http://article.gmane.org/gmane.comp.encryption.general/12588
/match=bitcoin+peer+to+electronic+cash+system.

42. "How DigiCash Blew Everything," *NEXT*, January 1999,
http://cryptome.org/jya/digicrash.htm.

43. Average Number of Transactions per Block, Blockchain Info,
December 2015, https://blockchain.info/charts/n-transactions-
per-block.

44. Robert Churchhouse, *Codes and Ciphers: Julius Caesar, the Enigma,
and the Internet* (Cambridge: Cambridge University Press, 2002),
92.

45. Simon Singh, *The Code Book: The Secret History of Codes and
Codebreaking*, (London: Fourth Estate, 2000), 120.

46. NIST, "Advanced Encryption Standard (AES)," FIPS
Publication 197, (June 2008),
http://csrc.nist.gov/publications/fips/fips197/fips-197.pdf.

47. NIST, "Recommendation for the Triple Data Encryption
Algorithm (TDEA) Block Cipher," NIST Special Publication
800-67, (Revised January 2012),
http://csrc.nist.gov/publications/nistpubs/800-67-Rev1/SP-
800-67-Rev1.pdf.

48. American National Standards Institute, ANSI X9.62:2005,
"Public-Key Cryptography for the Financial Services Industry,
The Elliptic Curve Digital Signature Algorithm (ECDSA),"
http://webstore.ansi.org/RecordDetail.aspx?sku=ANSI+X9.62
%3A2005.

49. Bruce Schneier, *Applied Cryptography: Protocols, Algorithms, and
Source Code in C*, Second Edition (Hoboken, NJ: Wiley, 1996), 35,
355.

50. Carolyn Watters, *Dictionary of Information Science and Technology*,
(Academic Press, 1992).

51. Slava Gomzin, *Hacking Point of Sale: Payment Application Secrets,
Threats, and Solution*, (Hoboken, NJ: Wiley, 2014), 101.

52. FIPS, "Secure Hash Standard (SHS)", FIPS Publication 180-4, (March 2012), http://csrc.nist.gov/publications/fips/fips180-4/fips-180-4.pdf.

53. The hash function RIPEMD-160, (2012), http://homes.esat.kuleuven.be/~bosselae/ripemd160.html.

54. R.L. Rivest, A. Shamir, L. Adleman, "A Method for Obtaining Digital Signatures and Public-Key Cryptosystems," Communications of the ACM, vol. 21, issue 2, (1978), http://dl.acm.org/citation.cfm?id=359342&dl=ACM.

55. PKCS-1, RSA Security, 2015, ftp://ftp.rsasecurity.com/pub/pkcs/pkcs-1/pkcs-1v2-1.asn.

56. Euler's Totient Calculator, 2015, http://www.javascripter.net/math/calculators/eulertotientfuncti on.htm.

57. Modular inversion, 2015, http://www.cs.princeton.edu/~dsri/modular-inversion-answer.php?n=7&p=120.

58. Modular Multiplicative Inverse, 2015, http://planetcalc.com/3311/.

59. Secp256k1, Bitcoin wiki, March 2015, https://en.bitcoin.it/wiki/Secp256k1.

60. Elliptic Curve Cryptography (ECC), Certicom, 2015, https://www.certicom.com/index.php/ecc.

61. Neal Koblitz, "Elliptic Curve Cryptosystems," Mathematics and Computations, vol. 48, issue 177, (1987), http://www.ams.org/journals/mcom/1987-48-177/S0025-5718-1987-0866109-5.

62. Taher ElGamal, "A Public-Key Cryptosystem and a Signature Scheme Based on Discrete Logarithms", *Advances in Cryptology: Proceedings of CRYPTO 84*, (New York: Springer, 1985), http://link.springer.com/chapter/10.1007%2F3-540-39568-7_2.

Index

www.ingramcontent.com/pod-product-compliance
Lightning Source LLC
Chambersburg PA
CBHW071141050326
40690CB00008B/1528